Music Apps for Musicians and Music Teachers

Music Apps for Musicians and Music Teachers

Elizabeth C. Axford

ROWMAN & LITTLEFIELD
Lanham • Boulder • New York • London

Published by Rowman & Littlefield
A wholly owned subsidiary of The Rowman & Littlefield Publishing Group, Inc.
4501 Forbes Boulevard, Suite 200, Lanham, Maryland 20706
www.rowman.com

Unit A, Whitacre Mews, 26-34 Stannary Street, London SE11 4AB

British Library Cataloguing in Publication Information Available

Library of Congress Cataloging-in-Publication Data

Axford, Elizabeth C., 1958– author.
  Music apps for musicians and music teachers / Elizabeth C. Axford.
      pages cm
  Includes bibliographical references.
  ISBN 978-1-4422-3277-8 (pbk.) — ISBN 978-1-4422-3278-5 (ebook)
1. Music—Computer programs. 2. Music—Instruction and study—Computer programs.
3. Mobile apps. I. Title.
  ML74.3.A96 2015
  780'.0285—dc23

                                                                    2014036357

To my family, friends, colleagues, teachers, and students,
to whom I owe my deepest gratitude for your love, support, and
encouragement through the years.

Contents

Acknowledgments

$\mathcal{I}$ would like to thank my immediate family for their love and support, including my parents, Dr. Roy A. Axford and Anne R. Axford; my brothers, Dr. Roy A. Axford, Jr. and Dr. Trevor C. C. Axford; my sisters-in-law, Cyndi Axford and Sarah Axford; my nephews, Noah and Charlie Axford; my nieces, Madison and Mackenzie Axford; and my godparents, Fleur and Jay Chandler. Thanks for always loving, supporting, and believing in me. Thanks to my editor at Rowman & Littlefield, Bennett Graff, for understanding my vision. Thanks to my assistants, Kathy Alward, Dee Crowell, and Paul Heimgaertner, for your help. Thanks to my Grammy and travel buddies, Bridget Brigitte McDonald, Jean-Pierre Prieur, Glenn Turanza, Donna Fant, and Jamie Cutler, for the fun and adventure. Thanks to all my friends and colleagues, the hundreds of musicians and songwriters I have known and worked with through the years, and the hundreds of students whom I have been blessed to teach. You inspire me more than you'll ever know.

Introduction

$\mathcal{C}$reating, performing, responding, and connecting to music are processes in which we all engage. Singing, playing instruments, moving to music, and creating music enable us to enjoy music and to acquire musical skills and knowledge. We learn by doing. Learning to read and notate music gives us skills with which we can explore music independently and with others. Composing, performing, listening to, analyzing, and evaluating music are important building blocks for learning music. To participate fully in a diverse, global society, we must understand our own historical and cultural heritage as well as that of others within and outside our communities. Because music is a basic expression of human culture, everyone should have access to a comprehensive, balanced, and sequential program of music study.

Learning to play an instrument or sing as well as learning about music plays an important role in a student's development, not just as a means of learning performance repertoire, but also as a means of nurturing the student. Students have different strengths, interests, and learning styles, all of which are considered when choosing the appropriate curriculum and lesson materials for studying music. Aspiring musicians can develop musical artistry and personal expression in addition to the achievement of musical skills and knowledge. Our responsibilities as music teachers include helping students develop an appreciation for all styles of music; providing musical challenges, music education, and performance opportunities; and preparing young students for college music departments as music majors or nonmajors and/or professional performing and teaching careers.

The National Core Arts Standards help to structure the measurable objectives for learning music in grades K through 12. The following chapters are based on the new standards that can be reviewed at http://musiced.nafme .org/musicstandards/ and http://www.nationalartsstandards.org.

Music Apps for Musicians and Music Teachers includes annotations of 1,090 iOS and Android music apps. These music apps can be used for many purposes, including learning music theory and music history, singing and ear training, playing an instrument, listening to music, composing music, and recording. Many of the apps listed overlap in purpose and function. I selected these apps after conducting a yearlong scavenger hunt consisting of numerous searches in the Apple iTunes App Store, Google Play, and on the Internet. The intent here is to show some representative examples of music apps that are helpful in the learning and enjoyment of music.

I have included a link to the app developer's website when one was available. The app developer's website shows updates and new apps as well as links to the Apple iTunes App Store and/or Google Play. It will also show if an app that was once available only for iOS is now also available for Android, for example. Visiting the app developer's website is a good way to see what other apps they offer as well as where to go for support should there be any bugs in the current version. As with music software programs, music apps come and go. The apps listed here were available at the time this book went to press.

Compared to many of the educational music software programs once available, music apps are far more affordable and accessible. In the past, when I would suggest that a student or parent purchase a music software program, they often weren't sure where to purchase it, which version to buy, how to install it, or how to use it. These factors, along with the fact that many of the music software programs were quite expensive, made it difficult to have students use them at home on their own time. Most of my use of educational music software programs with students was in my own studio during their lessons. With music apps, students can easily and quickly download a multitude of learning tools at affordable prices and use them virtually anywhere provided they have a smartphone or a tablet with either the iOS or Android operating system.

This book is not specifically a "how to" book on how to use iOS or Android devices. I have included a bibliography and a list of websites at the end with some helpful resources that further serve that purpose. I am simply listing some of the many music apps available in an organized and approachable fashion, and no product endorsements are intended. It's up to the reader to decide which apps to purchase and use. Some of the apps are free with additional features available upon purchase of the full version and/or a subscription service. I have not included any prices, as these are likely to change.

The numerous music apps listed here can be used as tools to assist in achieving the National Standards for Music Education. Determining the curriculum and the specific instructional activities necessary to achieve the standards is the responsibility of states, local school districts, and individual teachers. This reference book serves as a guide, as did my previous book, *Song Sheets to Software—A Guide to Print Music, Software, Instructional Media, and Web Sites for Musicians* (Scarecrow Press, 2009).

It's easy enough to do your own search for music apps that suit your interests and needs. I hope you find this book helpful and that you have as much fun discovering the world of music apps as I have had—*bonne aventure!*

Part I

CREATING MUSIC

· *1* ·

Reading and Notating Music Apps

MUSIC THEORY APPS

4 *Music Rooms* (http://t-c.gr) Educational music theory knowledge and entertainment game; in the four rooms of the house the user can access the four families of musical instruments: strings, woodwinds, percussion, and brass; (iOS 5.0 or later).

Circle of 5ths Master HD (http://eunjae7.wordpress.com/apps/) App unlocks complete enharmonic key signatures that help give an understanding of the circle of fifths and other related issues; works well with the other two versions of Circle of 5ths with multitasking between the three apps; explore and discover hidden musical keys; series is a comprehensive library for all musicians, composers, music teachers, lecturers, professors, students at all levels, and music enthusiasts; includes various musical components such as key signatures, intervals, scales, triads, and more; designed to help develop aural perception including relative pitch and perfect/absolute pitch; use as a reference as well as a study and teaching guide for training of the ear; learn, develop, and maintain musical knowledge and sharpen the ear; (iOS 7.0 or later).

Circle Theory (http://www.artsiness.com/Artsiness/Circle_Theory.html) Tool to assist seeing the relationship between notes; key signatures for major and minor scales; assists in transposing; guide to note intervals per musical mode; VGA adapter can be used to view the circle on an external screen or projector; tap or drag on the outer ring to change the tonic note and key signature; double-tap the outer ring to cycle through the inner ring displays; double-tap the center of the circle to cycle through the clefs; when the inner ring is displaying a musical mode, drag the inner ring to select a new mode; when the inner ring is displaying the transposition assistant, drag the inner ring to select a new starting note; (iOS 7.0 or later).

ClefTutor (http://www.cleftutor.com) Learn how to read music, identify music notes, memorize every key signature, and study music theory; developed by a music teacher veteran; interactive quiz game; practice sight-reading; for all musical instruments including piano, keyboard, guitar, bass, violin, viola, cello, flute, etc.; for all levels of musicians; music note reading app includes treble, bass, alto, and tenor clefs; supports note names, movable and fixed Solfège, and scale degrees; improves musical memory, speed, and accuracy; memorize the notes on the staff in all keys and common clefs; multiple quiz levels; answer with buttons, a one-octave keyboard, or an eighty-eight key piano; random clefs and key signatures; master the ledger lines; custom ledger line settings; custom options let user choose any combination of clefs, note ranges, and more; save settings and statistics for up to four users; track practice time and progress with statistics; for phones and tablets; works with Adobe Air; (Android 2.2 and up).

Echometry (http://echometry.com/content/ipad-iphone-app) Teaching system that shows how simple the different chords, scales, and modes are on the Echometric wheel; sound cell is made up of twelve consecutive sounds usually referred to as an octave or chromatic scale; ear-training and music–writing tool, including both a-b-c and do-re-mi teaching systems in all seven octaves; signature wheel for all the Greek modes; instructional tool that simplifies learning music by utilizing colors and patterns arranged in a circle with twelve dots, one for each note; student has an immediate visual connection with the notes in a scale while at the same time seeing that scale harmonized with a set of color-coded dots to show which chords should be major, minor, diminished, or augmented in any given key; (iOS 4.3 or later).

Glossary of Music (Wan Fong Lam) Contains more than five thousand music-related terms and jargon; for musicians as well as students who are studying music and those who want to learn more about the history and theory of music; (iOS 3.0 or later).

Music Learning Lab (http://www.musiclearninglabapp.com) Covers key skills that serve as a foundation for music learning; in the Learn area, user learns new musical skills that provide an essential foundation for music education; completing lessons unlocks new instruments and backing bands to use in the Create area; use new skills to create musical pieces and see them performed by animated characters; choose pitches, change instruments, add longer melodies and more complex rhythms, and choose a backing band to play along; more instruments can be earned; in the Play area, user practices skills learned in the lessons with three mini-games; completing the games unlocks more instruments and backing bands; see what a learner has accomplished in the Trophies area; see trophies that represent everything a learner has achieved and watch animations that demonstrate accomplishments; explore the app to earn all fifteen instruments, nine backing bands, and thirty-two trophies; no ads; (iOS 5.0 or later).

Music Resources (http://rondostudios.com/musicresources.html) More than two hundred music educational documents for the classroom and studio; includes blank notation and tab paper, flashcards, worksheets, chord charts, instrument diagrams, games and puzzles, reference materials, instrument exercises, sheet music, and practice exams; Universal app; all content has been created in the standard Portable Document Format (PDF) and optimized for printing on A4 sheets of paper; ready to use with Air Print; send handouts as e-mail attachments to students and fellow teachers; open documents within other supported apps on device for viewing and editing; make requests for new content; no Internet connection is required; (iOS 7.0 or later).

Music Theory and Practice by Musicopoulos (http://www.musicopoulos.com/ios-iphone-ipad-app-reviews) Combines structured music theory lessons with matching exercises; subjects include: Introduction to Intervals, Introduction to Scales, Circle of Fourths and Fifths, Key Signatures, Diatonic Intervals, Major Chords, Diatonic Intervals Continued, Relative Minor Scale, Inverted Major Chords, Chords Continued, Descending Intervals, Chromatic Intervals, Diatonic Chords, Harmonic Minor Scale, Melodic Minor Scale, Inverted Intervals, Major 7th Chords, Minor 7th Chords, Dominant 7th Chords, Major 6th Chords, Pentatonic Scale, Minor Pentatonic Scale, and Modes; comes with a default piano sample that will help develop ears and musical knowledge; other instrument choices include: piano, guitar, bass guitar, violin, viola, and cello; understanding music theory helps user become a better musician; (iOS 6.1 or later).

Music Theory Basics—Patrick Q. Kelly (http://patrickqkelly.com/index.php/ipad/music-theory-basics) Suite of modules to help test basic music skills; covers notes, keys, intervals, and chords in all four clefs (treble, alto, tenor, and bass); set how many questions there are in the quiz (one to fifty); set the option to only grade the questions answered; work in ink on paper or chalk on a chalkboard; timer to record how long it takes to finish the quiz; *Notes—Ear Training:* hide the staff or keyboard showing the requested pitch; set the quiz note range from zero to nine ledger lines above and below the selected staff; up to three tries to get the correct answer; notate the pitch that is played on the piano; play the written note using the piano keyboard; *Modes and Key Signatures:* test knowledge of modes as well as major and minor keys; set a maximum number of sharps or flats to work on; *Intervals—Compound/Simple:* set how the question is asked; play the interval harmonically, melodically moving up or melodically moving down; option to have the intervals given with random root notes within the set range or set what the root will be for the interval; intervals covered are diminished, major, minor, perfect, and augmented; *Chords:* set how the question is asked; play the chord normally, rolling up or rolling down; set a random root movement or set the root during the quiz; all seventh types are covered; test without the sound for work on identifying the chords by sight alone; *Rhythms:* polyrhythms are lines of rhythms sounding simultaneously; select the specific combination of note values to include in the test rhythms: whole, half, quarter, eighth, dotted half, dotted quarter, triplet half, triplet quarter, and triplet eighths; set the length from four to thirty-two beats of the test rhythm; include or exclude grading durations of note values; include or exclude using rests in the rhythms; select from one to three pads/pitches notated above, below, or on the lines; select a range of tempo up to 120 bpm; listen to the rhythms before playing; use the timer to set an amount of time to work on a specific skill; enter the student's name either in the module itself or in each of the module setup preferences pages; all the modules allow the user to set the number of tries to answer each question; enter the student's name and it passes between the modules, recording the scores in each one as well as the final combined grades tab; teacher can see how students have done in each area; access can be locked and unlocked within the app using a password set in the iPad's main settings app; (iOS 5.0 or later).

Music Theory by Gregg Fine (www.macprovideo.com/about/trainers/greg gfine) Apps and videos include the following: Music Theory 101—Melody: learn the theory and art of melody writing; Music Theory 102—Harmony Chords: learn the theory behind creating chords; Music Theory 103—Rhythm: explores the nature of rhythm in music; Music Theory 104—Song Form: understanding song form is essential to songwriting; Music Theory 105—Basslines: break down bass lines and learn how to create grooves; Music Theory 106—Building Chord Progressions: thirty-tutorial course on how to create chord progressions; (iOS 4.2 or later).

Music Theory for Beginners (http://www.musicroom.com/music-trivia-apps) Includes twenty-two lessons that cover all the major music theory topics from reading notes and rhythm to basic harmony; take the interactive quiz and get an achievement badge for each topic; progress onto the next lesson; contains more than two thousand questions; music theory tool; (iOS 4.0 or later).

Music Theory for Dummies (http://support.inkling.com/hc/en-us) Interactive edition on Inkling; includes everything from the print book plus audio clips, live links to other sections and the web, a search feature, and more; optimized for each device; Universal app; listen to embedded music clips and follow along with the music without turning a page; go directly to websites mentioned anywhere in the book; search anything and Inkling looks through the whole book as well as Google and Wikipedia; Notebook for bookmarks and highlights; (iOS 5.1 or later).

Music Theory Lessons (http://www.nadstech.com/download-free-apps/music-apps) Includes: Staff, Clefs, and Ledgers; Note Duration; Measures and Time Signature; Rest Duration; Dots and Ties; Steps and Accidentals; Simple and Compound Meter; Odd Meter; The Major Scale; The Minor Scales; Scale Degrees; Key Signatures; Key Signature Calculation; Generic Intervals; Specific Intervals; Writing Intervals; (Android 1.6 and up).

Music Theory Pro (http://www.musictheorypro.net) App challenges user by testing knowledge; covers note names, key signatures, intervals, chords, and ear training; adjust level of difficulty by tapping on gear button and customizing quiz question topics; test is timed; tap the question mark button to review concepts; posts scores; identify by ear tempos, intervals, chords, and scales; requires at least some music theory knowledge; (iOS 6.0 or later).

Music Theory Video Tutor (http://www.ipreppress.com/Page1/musiced .htm) Upon completing the course, user will be able to read and write music; includes fifty-three music theory lessons with exercises that have a cursor in sync with the music as it is played; The Language of Music Part 1 and The Language of Music Part 2; (iOS 4.3 or later; Android 2.2 and up).

Music Theory with AUDIO (https://www.brainscape.com/market/know ledge_junkie/music_theory/music_theory_crash_course) Develop listening skills and tone control; identify written chords; increase knowledge of music theory and classical composers; music theory flashcards; includes fifteen AUDIO decks: Beginning Chord Theory, Music Symbols, Key Signatures, Name That Chord, Closely Related Keys, Non-Chord Tones, The Seven Modes, Harmonic Cadences, Chord Ear Trainer, Interval Ear Trainer, Beginner Rhythm Trainer, Advanced Melody Ear Trainer, Composer Trivia, Common Concert Instruments, and World Performance Halls Trivia; (iOS 7.0 or later).

Musictionary Music Dictionary (http://www.rodskagg.se/portfolio/music tionary/) Music dictionary; more than five hundred words with descriptions and images; bookmark favorite words; search function to find words or browse entire dictionary; (iOS 6.0 or later).

Nota HD (http://notaapp.com/) Tools for musicians at any level; piano chord and scale browser; piano and staff note locator; note quiz and a reference library with more than one hundred symbols; for beginners, app covers the basics of music notation with a four-octave piano that displays the notes on a staff, a full screen landscape mode piano for practicing, and an interactive notes quiz to test knowledge of notes; for advanced users, the scales browser has a comprehensive list of common and exotic scales; will show the scale, play it, and display the notes, intervals, and half-steps; chords browser makes it easy to find a chord and play it on any key or invert it; set the notation to strict or simplified and set the root to sharp or flat; consult the Circle of Fifths in the Reference section with a comprehensive reference of music notation; note quiz tests knowledge of notes on the staff; difficulty level can be set to easy with thirty-four notes or advanced with eighty-two notes that include sharps and flats; Music Brainiac Reference Library with more than one hundred items that include: Accents and Accidentals, Lines, Breaks and Clefs, Key Signatures, Chords, Circle of Fifths, Dynamics, Note Relationships, Notes, Notes and Duration, Repetition and Codas, Rest and Durations, Time Signatures, Browse with a Flick, Parlez Vous Solfège; in fixed Do Solfège mode, notes are shown in their Solfège syllables do, re, mi, fa, sol, la, ti, do; not a music tutor nor should it replace an actual teacher; tool for anyone doing music, beginner or experienced; (iOS 4.0 or later).

Octavian Pro (http://www.bitnotic.com/octavian.html) Music theory app featuring more than 550 scales and more than one hundred chords with options for viewing, listening, and cross-referencing; reference, learning, and teaching tool for all musicians, students, composers, and songwriters; includes scales from around the world and through history; view any scale with any root note, in any mode, and examine the chords and harmony within the scale; chords including sevenths, elevenths, ninths, thirteenths; invert chords and browse scales that include the chord; compose twelve-step chord progressions with inversions, transposition, and added sevenths for each step; enter notes on the keyboard to find the scales and chords that contain the notes; configure the note labels to show notes, intervals, degrees, steps, or Solfège; choose tempo, octave, and repeat options for the piano sounds to suit ear-training, practice, or for hearing new harmonies; PDF user manual and video available at website; (iOS 4.3 or later).

Tenuto (http://www.musictheory.net) Twelve customizable exercises designed to enhance musicality; four musical calculators for accidentals, intervals, chords, and analysis symbols; staff-based exercises include: Note Identification—identify the displayed note; Key Signature Identification—identify the displayed key signature; Interval Identification—identify the displayed interval; Chord Identification—identify the displayed chord; Keyboard-based Exercises: Keyboard Identification—identify the note name of the piano key; Keyboard Reverse Identification—identify the note by pressing a piano key; Keyboard Interval Identification—identify the interval of the highlighted piano keys; Keyboard Chord Identification—identify the chord of the highlighted piano keys; Fretboard-based Exercises: Fretboard Identification—identify the note name of the marked position; Ear Training Exercises: Interval Ear Training—listen to the played interval and identify its type; Scale Ear Training—listen to the played scale and identify its type; Chord Ear Training—listen to the played chord and identify its type; Calculators: Accidental Calculator—display the accidental for a note and key; Interval Calculator—display the interval for a note, type, and key; Chord Calculator—display the chord for a note, type, and key; Analysis Calculator—display the chord for a symbol and key; (iOS 5.1 or later).

Theory Lessons (http://www.musictheory.net) Features thirty-nine music theory lessons from musictheory.net in their original animated versions; complete lesson list: The Staff, Clefs, and Ledger Lines; Note Duration; Measures and Time Signature; Rest Duration; Dots and Ties; Steps and Accidentals; Simple and Compound Meter; Odd Meter; The Major Scale; The Minor Scales; Scale Degrees; Key Signatures; Key Signature Calculation; Generic Intervals; Specific Intervals; Writing Intervals; Interval Inversion; Introduction to Chords; Triad Inversion; Seventh Chords; More Seventh Chords; Seventh Chord Inversion; Diatonic Triads; Roman Numeral Analysis: Triads; Diatonic Seventh Chords; Roman Numeral Analysis: Seventh Chords; Composing with Minor Scales; Voicing Chords; Analysis: "O Canada"; Nonharmonic Tones; Phrases and Cadences; Circle Progressions; Common Chord Progressions; Triads in First Inversion; Triads in Second Inversion; Analysis: "Auld Lang Syne"; Building Neapolitan Chords; Using Neapolitan Chords; Analysis: "Moonlight Sonata"; (iOS 5.1 or later).

Treble Clef Kids (http://www.christianlarsenmusic.com/Christian_Larsen/ Treble_Clef_Kids.html) Learn to read music and master the basics of music theory with the Treble Clef Kids lesson series; tutorial apps for learning and practicing important sight-reading and musicianship skills; fun, interactive, and educational way for kids and adults to learn the basics of music theory; exposure to music correlates to better performance in school and in life; simple and effective music theory application; learn basic musicianship skills in a logical, guided way, progressively building through the applications; practice and performance modes; interactive playing of the piano encourages students to seek the correct answer through the sound of applause while providing quirky sounds for wrong answers without penalty; (iOS 4.2 or later).

Virtuozzy Rhythm Builder (http://www.virtuozzy.com/virtuozzy/HOME .html) Explore the world of rhythm; fun and intuitive music bar building game for young music students ages four to twelve; (Android 2.2 and up).

Wolfram Music Theory Course (http://products.wolframalpha.com/course assistants/) Music reference from Wolfram; learn about notes, intervals, scales, and chords and hear what they sound like; choose from both common scales and hundreds of more advanced scales; explore triads and basic major, minor, and seventh chords; input up to four chords and hear the progression; learn how to identify music intervals by name and what they sound like; find interval inversions for every interval type; reference musical terms in the abbreviated music dictionary; (iOS 5.0 or later).

NOTE IDENTIFICATION AND SIGHT-READING APPS

Blue Note Music Flash Cards (http://bluenotemusicflashcards.appspot.com) Learn musical notes by memorizing one set of notes at a time via random repetition; begin at level one and build musical note recognition while advancing through all levels until all notes have been mastered; (iOS 6.0 or later).

ChordSense—Pitches (http://www.guitarcreativity.net/Home.html) Introduces pitch names and their locations on the staff; for beginning musicians or those who are just learning to read music; user can choose between bass clef and treble clef; learn the notes on the lines and spaces of the staff and on the ledger lines above and below the staff; introduction to other apps in the ChordSense series including *Basic Intervals* and *Specific Intervals*; (iOS 4.3 or later).

DoReMemory (http://www.gadjetts.com/DoReMemory) Music educational app created to teach early pianists where notes belong on the treble and bass clef staffs; user is given the letter name of a note and then taps where they think that note should be on the staff; when correct, the note shows up, the correct pitch plays, and stars twinkle; when user does not tap the correct spot, an arrow directs them to the right location; no penalty for incorrect answers; try again or move on to the next note; no timer; includes treble notes from middle C to a high G and bass notes from middle C to a low F; Middle C is the only ledger line note included; to switch between clefs, tap the treble or bass clef symbol in the lower corners of the screen; switch from letter note names to the "fixed do" Solfége names; (iOS 4.3 or later; Android 1.6 and up).

DoReRhythm (http://www.gadjetts.com/DoReRhythm) Choose a time signature and drag notes onto the music staff; if note counts match the time signature, measure will play; practice counting with whole notes, dotted half notes, half notes, quarter notes, and eighth notes; in easy mode, focus on the whole notes, half notes, and quarter notes; in easy mode, time signature choices are three/four time and four/four time; hard mode adds in six/eight time; settings button provides choices and gives some tips for those just learning to count music; understanding rhythm is important for good sight-reading skills; (iOS 3.2 or later).

Flashnote Derby—Musical Note Flashcards! (http://flashnotederbyapp.com) Learn music notes and practice memorizing them; each race is a music drill where the user identifies different notes as they appear on each flashcard; answering quickly and correctly causes horse to gain ground; incorrect answers will cause horse to fall behind; at the end of each race, user can review the notes missed and see the correct answers; game is customizable to allow for any desired notes in the treble clef or bass clef to be included; select exactly which notes are drilled; amount of time given for each music flashcard can be increased or decreased depending upon the user's age and experience; drills can start out simply and gradually be made more challenging until the full staff is mastered including up to two ledger lines above the treble staff and two ledger lines below the bass staff; the more difficult the settings, the more points are awarded; high score is tracked so that players always have a goal; more than a dozen instructional video lessons are accessible from inside the app; children with no prior music instruction or note reading experience can start learning about the music alphabet, the staff, and how to identify different notes; provides a motivational tool for music teachers to use with students; for piano students, band and orchestra students, singers, and guitar players; (iOS 5.0 or later; Android 2.3.3 and up).

Freddie the Frog Pro (http://www.freddiethefrog.com/app.php) A music education app correlating with the Freddie the Frog books; introduces the written language of music; Freddie and his best friend, Eli the Elephant, love to go on adventures in the Sea of Music, unlocking the world of music as they explore Treble Clef Island, Tempo Island, Crater Island, and more; Freddie's encounter with the Bass Clef Monster introduces the bass clef notes; review music concepts learned through the stories; children take advantage of kinesthetic learning as they paint illustrations from the books, challenge themselves with note matching games, play note name games, and create music with rhythm games; (iOS 7.0 or later).

Hear It, Note It!—The Aural iQ Game (http://mynotegames.com) Music puzzle game for beginners to experts; listen to the notes, remember them, and then write them with the drag-and-drop interface; learn and test skills with note, rhythm, pitch, and scale puzzles; need a rudimentary knowledge of musical notation to get started; may need help from a music teacher to cover the more difficult levels; understand musical notation and phrases; (iOS 4.0 or later).

mpt Note Speller (Supply Street Inc.) Introduction to music notation; learn about the basic terms used in musical notation and master the ability to name and locate all the notes on the treble and bass clefs of the grand staff; test skills with ledger lines; (iOS 4.0 or later).

Music Notes (http://www.foriero.com/en/pages/music-notes-page.php) Exercise note recognition skill; recognize music notes as they appear in music; includes Train Mode and Play Mode where user has sixty seconds to recognize as many notes as possible; (iOS 5.1 or later).

Music Notes Flashcards (Totally Integrated Mobile LLC) Touch the screen to see a music note; tap the screen to see the answer; customize which notes user is quizzed on; select only treble or bass notes and line or space notes; beginner and advanced mode; (iOS 3.0 or later).

Music Reading Essentials (http://apricotpublishing.com/support.aspx) Learn and practice sheet music reading, musical signs and symbols, note names, rhythms, intervals, and more; designed with the help of professional piano teachers; for those who are learning to play piano or other musical instruments as well as anyone who needs to practice sheet music reading; includes four sets of flash cards divided into stages; choose either one or a combination of any of the four sets; shuffling algorithm shows cards answered incorrectly more often; displays results; (iOS 5.0 or later).

Music Tones (http://www.foriero.com/en/pages/music-tones-page.php) Game is focused on reading music tones on the staff; teaches where to find notes on the music staff with either treble or bass clef; uses colors; (iOS 5.1 or later).

Music Tutor (http://www.jsplash.com/apps.html) Improve sight-reading skills; practice among treble, bass, or both clefs for a duration of one, five, or ten minutes; review mistakes and see progress; (iOS 7.0 or later).

Musical Me! HD (http://www.duckduckmoose.com/educational-iphone-itouch-apps-for-kids/musical-me/) Teaches notes, rhythm, and pitch; listen to notes, copy patterns, and recognize pitches; learn about rhythm, including short and long notes; play with a drum, cymbals, triangle, maracas, or egg shaker; learn to read music and create songs by moving notes on a staff; listen to children's songs featuring the violin and cello; for ages three to seven; train ear to hear different pitches; tap, drag, or hold the monsters to make them dance to the beat; (iOS 5.0 or later).

My Note Games (http://www.appatta.com) Suite of simple games to teach music theory and instrument mastery; listen to the music; write what is heard by using drag-and-drop interface; tap on the buttons to name the notes; forty-eight levels each for treble, bass, tenor, and alto clefs; first four notes of each clef are free; premium version includes the rest; recognizes the notes user plays on the instrument; first three notes of all instruments are free; premium version includes the rest of the notes for all the instruments plus Tap That Note; choose a lesson to play with each instrument; play eight phrases and get the best score; Play-A-Day uses computer-generated phrases; premium version includes lessons for all the instruments including saxophone, piano, guitar, descant/soprano recorder, treble recorder, flute, oboe, clarinet, bassoon, trumpet, vocals, and whistling; with Toonr The Tuner user can see the note playing displayed on the screen; designed to be used under the supervision of a parent or music teacher; uses note recognition technology; (iOS 4.0 or later).

Note Perfect (http://noteperfectapp.com/) Test and improve note reading against the clock; for all ages; gameplay in choice of four different clefs; play as beginner, intermediate, or virtuoso; compete with friends for the high score; (iOS 3.1.3 or later).

Note-A-Lator (http://www.electricpeelsoftware.com/note-a-lator-ios/) Improve music reading and theory skills; make and print customizable quizzes on Note Identification, Rhythm Identification, Key Signature Identification, Scale/Modes Identification, Chord Identification, and Interval Identification; (iOS 5.1 or later).

Notes for Little Composers (http://littlecomposers.com/apps.html) Interactive app that teaches new music students the note names: A B C D E F G; click on a note and listen; play a note guessing game; for children age six or older; (iOS 3.0 or later).

Notes!—Learn to Read Music (http://visionsencoded.com/fun-iphone-apps/) Basic flashcard app to learn the letter names for notes on the bass and treble clefs including "F" on bass through to "G" on treble; key letters can be set to: (1) C D E F G A B (2) C D E F G A H (3) Do Re Mi Fa Sol La Ti; displays a random note on the musical staff; tap the correct letter/key; MIDI enabled; use with Apple Core MIDI compliant devices; change to either treble or bass clef by tapping the clef; toggle visual effects by tapping trophy icon; practice only specific notes; (iOS 4.3 or later).

NoteWorks (http://www.doremiworld.com/Games/NoteWorks) Musical game designed to teach note recognition and improve sight-reading skills; Hungry Munchy is eager to swallow elusive blue notes; goal is to help Munchy catch each note as quickly as possible; app is both educational and entertaining with an intuitive and visually appealing interface; created for players of all ages, families with multiple players, and music teachers with multiple students; game design features include a funny animated character, user-friendly menus, graphics, and sound effects; note reading options include on-screen piano keyboard input, "ABC" note naming, "Do Re Mi" note naming, and the treble clef; students learn by playing a game and having fun; in practice mode, students are helped with hints; in challenge mode, students play for points and stars; students learn to recognize notes; game speed is adjustable; (iOS 5.0 or later).

Notezart (http://notezart.noveogroup.com) Educational game to learn music notes and improve sight-reading and musical ear; select treble or bass clefs with different levels; sixty-eight levels in total; in each level, there are different pieces of music with three educational mode options: Learn Mode, Arcade Mode, and Hear Mode; music notes flow on the screen with the piano sounds; players have to hit the correct music note on the piano keys below the staff ledger; if it is the right music note, the note turns green; in the Learn Mode, if the player hits the wrong music note or doesn't hit any piano keys, the correct note appears on the pop-up screen; in the Arcade Mode, the speed of the game increases; in the Hear Mode, the player hears the piano sound of the note but does not see it on the screen; (Android 2.3 and up).

Sonja and *Sonja Classroom* (http://www.choirpro.com) Read and write music; simplified and intuitive music notation system that lets user follow a score; learn and practice any part by lowering the volume or muting any other part, slowing down playback, or selecting part of the song for looped playback; experiment with the music staff, music notes, and the metronome, while learning the concepts of pitch, note duration, rests, flats and sharps; create and notate music; import and edit MIDI files; import audio files; record audio files of instrumental accompaniments or singing; export MIDI files for further editing and publication; export Sonja-format files that combine music notes and audio tracks; (iOS 6.1 or later).

SCALES AND CHORDS APPS

Advanced Chord Progressions (http://www.walking-library.com) More than 320 chord progressions and sound samples; chords progressions are modifiable; lists of chord progressions that work; an important goal of songwriting is to create something that listeners will remember, and they won't remember songs that use too many chords, melodies, or words; (iOS 3.2 or later).

Chord Detector (http://www.chord-detector.com/wordpress/apps/chord-detector/) For musicians of any ability; learn to play the tracks in iPod music library by automatically detecting the chords within the track; can recognize many different chord types including major, minor, augmented, diminished, diminished seventh, dominant seventh, minor seventh, and major seventh; option to view the detected chords in two different forms, scrolling and tabular; in both views user is able to play the track; chords will move or be highlighted as the song is playing; can view the chords of a track while analysis is going on; queue up tracks to be processed; controls on the chord view screens include: Play/Pause to play or pause the track; Rewind 30s to rewind the track by thirty seconds; Forward 30s to forward the track by thirty seconds; Tempo Adj. to slow down the track without affecting the pitch by a variable amount down to one quarter of its original speed; Capo Adj. to transpose the chords according to which fret capo is on; (iOS 5.0 or later).

Chord Explorer (http://www.mangomoo.com/chordexplorer/) Full eighty-eight key piano keyboard with assignable chord pads; MIDI controller; experiment with harmony and chord progressions; interval, chord, and scale reference; ear-training tool; store note combinations across twenty-four pads; play the pads and keyboard together; connect to a synth, sampler, or sequencer; identify notes, intervals, chords, and scales from a comprehensive library; enhance understanding of harmony; create ear training tests; share projects; (iOS 6.0 or later).

Chord NOTE (http://therootage.com/iphone/chord-note-en/) Songwriter's assistant tool; find chords to match melody; find song's key; (iOS 5.0 or later).

Chord Walker (http://cleverlevers.com/chordwalker/) Training tool for musicians who need to walk through chord changes; written by a bass player to practice Real Book sight-reading; goal is to master chord navigation on the instrument, reflecting the user's chosen style, personal taste, and tempo; one-tap generation of musically valid chord progressions; root name and chord character both shown; select starting key; select modulation interval: perfect 4ths or 5ths; choice of flat or sharp accidentals; save any progression as a favorite; (iOS 4.0 or later).

ChordGen (http://www.by2labs.com) Chord progression generator; generate chord progressions with the root note of choice; includes some of the popular chord progression sequences in all the modern scales and modes; get chord progressions for piano, acoustic guitar, and power chords; teaches user how to play each of the listed chords in the selected sequence for the particular musical instrument; build own chord progressions with a step by step process; save work; basic level songwriting application; plays the progressions in a loop and can be used as a backing-track tool; jam on the backing track of choice; (iOS 2.2 or later; Android 2.3 and up).

Chordion (http://www.olympianoiseco.com/Chordion/) Make music on the go; choose chords with one hand and play melodies with the other; customization options include a synthesizer, scales, and more; supports MIDI; control other instruments; (iOS 5.1 or later).

ChordLab (http://ipad.rogame.com/pages/ChordLab.html) Chord spelling for all common chords found in fake books; music notation and textual representation of chord tones; chord analysis to correctly identify triads and seventh chords; alto, bass, tenor, and treble clefs; TAB and grand staff; all possible inversions; all standard voicings; all common open and alternate tunings presets for guitar; alternate fingerings for guitar; virtual piano and guitar instruments; chord tone labeling in instrument views; chord rating; automatic reduction to three or four voices; automatic root and fifth substitution; key signatures for all major modes; circle of fifths; (iOS 6.1 or later).

DodoChord (http://www.dodochord.com) Play chords using two fingers; features fast chords switching, quick inversions, octaves, and simple flat/sharp notes swapping; assures finding the exact chords, inversions, and octaves needed while playing songs, doing ear training, testing chord progressions, or composing; melody will keep flowing without being interrupted; helps user concentrate on the rhythm, improv feedbacks, and the audience's interaction; ready-to-play virtual instrument; (iOS 4.3 or later; Android 2.2 and up).

DoReMi 1-2-3: Music for Kids (http://creativitymobile.com/doremi123/) Learn major scales, scale placement, and pitch recognition; includes sixty-four characters: sparkling stars, mooing cows, and more; over forty-five different levels; eight familiar children's melodies designed for easy learning; song mode and free mode; create free mode recordings; free play includes color, number, Solfége, and animal sound mode; (iOS 4.3 or later; Android 2.2. and up).

Dragon Scales (http://dragonscalesapp.com) Learn and practice the major and minor scales; slay the dragon and steal his treasure; scales are an important part of a student's development and should be included as part of his or her study; learn about key signatures, the order of sharps and flats, and all of the major and minor scales; customizable; drill specific scales; choose from all major and minor keys, including harmonic, melodic, and natural minors; as abilities improve, increase the attack speed of the dragon and earn more points; over an hour of instructional videos are included in the app with no additional purchase; Internet connection required; learn about the circle of fifths, the relationship between major and minor keys, and the specific notes of any particular scale; students preparing for the Royal Conservatory of Music

Examinations (Canada) or the assessments of the Music Development Program (USA) will find that the specific scale requirements for each grade level are available as presets; a single tap will enable all of the scales that could be requested on an examination; (iOS 6.0 or later).

iHarmony (http://www.iharmonyapp.com) Complete collection of scales, chords, and harmonizations found in music; more than 1,500 entries and audio clips; choose Anglo-Saxon notation (C, D, E . . .), Italian notation (Do, Re, Mi . . .), German notation (C, D, . . . H), or French notation (Ut, Re, Mi . . .); (iOS 7.0 or later).

Inversion Invasion (http://jambots.com) Learn keyboard chords and inversions with a fun arcade style game; (iOS 4.0 or later).

KeyChord—Piano Chords/Scales (http://www.umito.nl/#KeyChord) Dictionary and reference app for piano chords and scales; keyboard to look up all chords and scales; includes a five-octave, scrollable, multi-touch mini piano with 128 different sounds from Mini Piano Pro; more than fifty chord types; more than one hundred scale types; note names; component intervals; reverse chord and scale lookup, including partial chords; inversions/slashed chords; staff view; favorite and recent chord types; hear every chord and scale; record songs as MIDI and set them as a ringtone; audio playback is enabled in all modes; companion for learning fake book or Real Book jazz piano; Internet connection is only used for OPT-IN statistic; (Android 2.1 and up).

Keys-Chords (http://salcarusodesign.com/Sal_Caruso_Design/Keys-Chords .html) Helps music students, musicians, and vocalists understand the relationship between keys and chords in the major keys; identifies the major, minor, and diminished chords in each of the major keys; removes some of the mystery in transposing chords from one key to another; user interface consists of two views: the Chord View and the Transpose View; the Chord View identifies chords in each of the major keys; the Transpose View will help musicians and vocalists who have to change keys; for any instrument that can play chords; useful aid for anyone who wants to understand the relationship between keys and chords; (iOS 6.0 or later).

MovableChord (http://studio-phiz.com/apps/) Plays basic chords on root notes in a diatonic scale; helps user hear the chord and the chord progression and learn each note; good tool for learning to read music with scale degrees; use for searching/learning chords for a tune, building solos from chord tones, or studying music theory; (iOS 6.1 or later).

MScales (http://salcarusodesign.com/Sal_Caruso_Design/MScales.html) Helps music students, singers, musicians, and music aficionados develop a better understanding of the relationship of the notes in each scale of the major keys; user interface consists of two views: the Scale View and the Transpose View; the Scale View displays the notes in the scale of the selected key; each note is referenced as Solfége names "Do, Re, Mi, Fa, Sol, La, Ti, Do" and the numbers one through seven; numbers are provided for counting and singing in intervals; change the notes in a song to a different key so that the instrument will match the vocal range of a singer or other musical instrument(s) in a band or ensemble; the Transpose View is useful for converting the notes in one key to the notes in a different key; (iOS 6.0 or later).

Music Theory—Chords in Keys (http://stuartbahn.com/apps/music-theory-chords-in-keys-android-ios-app/) Music theory app to learn musical keys; learn all the three- and four-note chords within all twelve major, natural minor, harmonic minor, and melodic minor scales; choose a scale to work on and the chord types; app will display a key and a Roman numeral chord number; for guitarists, pianists, brass players, wind players, strings players, and singers; created by guitarist and music educator Stuart Bahn to help music students improve their musical knowledge; pages summarize relevant theory for each scale and chord type; tests use Roman numerals in generic format for all scales; (iOS 6.1 or later; Android 1.5 and up).

Music Tool (http://spicyclam.com/musictool.html) Explore the world of music; large variety of scales from around the world; selecting a key, scale type, and mode provides the scale with chords; tap on any note of the scale, on a chord name, or on the keyboard to hear how it sounds; use portrait view to see the Chords screen; larger view makes it easy to experiment with scale and chord sounds; (iOS 4.3 or later).

Musician's Little Helper (http://markontheiron.com/musicians_little_helper.html) Pick a scale from a list of sixty-eight available; pressing the Random button will jumble up six individual notes relevant to the chosen scale to create something unique; notes are placed in spaces on the main screen so user can see them and play them; experiment with changing the order of the different notes and add more by dragging them from the keyboard into the free empty spaces, creating a longer series of notes to play back; make a loop by adding the loop icon into a free space; aids in learning scales by highlighting on a keyboard which notes belong to each scale; adjust the playback tempo using the slider; (iOS 4.0 or later).

Musix Pro (http://shiverware.com/musixpro/) Innovative app for performing music; isomorphic note layouts; shows built-in harmonic relationships between notes; provides a wide range of octaves, scales, keys, chords, and modes without limiting creativity; multiple built-in instruments; support for OSC and Core/Virtual MIDI; most existing synthesizers work with Musix Pro as a controller; Audiobus supported input; Virtual MIDI port; (iOS 5.1 or later).

Polychord (http://polychordapp.com) Musical instrument for experts and beginners alike; write songs, play music, record, and share; hit any of the chord circles to play a chord; slide fingers over the strum keys and play the notes; go from major to minor; switch keys; play a scale; turn on auto accompany and play along to drums, bass, and an arpeggiator that has a range of time signatures to choose from; turn on MIDI for any of the instruments and use it as a controller for audio programs; bitshift effects, including detailed controls for the drums; decimate and distort sounds with full control over bits and bitrate; (iOS 5.0 or later).

Pro Chords—Instant Inspiration (http://www.prochords.dk) App developed to write songs with professional chord progressions; offers a variety of options as to what chord could be played next after the sequence of chords already chosen during the writing process of a new song; suggestions are based on thousands of successful songs and may inspire user to find and choose new chords that they may not normally have thought of in a particular context; useful for songwriters looking to compose songs of professional standard; helpful for professionals who are looking for new paths and alternative ideas; more than eight thousand different progression patterns from the latest fifty years of rhythmic music; patterns are used to suggest possible movements at any given point in a song based on the previous sequence of chords; suggests many options starting with the most common alternatives; developed to inspire, not to do the songwriting for the user; intended as a source of inspiration to boost creative ideas; (iOS 6.0 or later).

Reverse Chord Finder Pro (http://www.reversechord.com) Helps songwriters, composers, musicians, and students match chord names to the notes they are playing; (iOS 5.0 or later).

Scale-Master (http://www.cal30.biz) Provides a means of practicing intonation; more than two hundred scales for banjo, bass, guitar, piano, and mandolin; virtual instruments are playable and can be shifted into any position; supports left-handed display and note overlays; MIDI support; history popover; global transpositions for horn players; chord matching and more; (iOS 7.0 or later).

Scales & Modes for iPad (http://www.smappsoft.com/scales---modes.html) Interactive visual and audio reference for the diatonic scales and modes that are fundamental in music theory; view major and minor modes on the musical staff; see fingerings for both keyboard and guitar; listen to the scales note by note, in any key; diatonic musical scales and modes are one of the fundamental concepts of Western music theory; mobile reference guide for viewing and listening to a large number of scales and modes, including all the modes of the major, natural minor, and harmonic minor scales; select a root note and a scale to display the notes of the scale on the musical staff; click Play Scale to listen; select the keyboard or guitar tabs to view fingering diagrams for the scale on those instruments; press the info button to see more detail about constructing the scale; (iOS 3.2 or later).

Scales Tutor (MUSIKEYS, LLC) Offers instruction for beginners with some musical knowledge; brush up reviews of major and minor key signatures and scales; recognize and enter all key signatures and scales on the staff; brief introduction to music staff notation; interactive tutorials; guided practice; practice for theory tests; (iOS 3.2 or later).

Suggester—Chord Progression Tool and Musical Scale Reference (http://www .mathieurouthier.com/suggester/) Tool to assist in the creation of songs and chord progressions; find chords that work together; build musical phrases that carry emotion through tension and release; touch a chord to hear how it sounds; press the play button to hear the chord progression sequentially and adjust the playback speed; work either forward or backward; pick a scale, then build a song from the chords that the app suggests; from the catalog, pick a set of chords and app will match scales; save experimentations for later reference; export by e-mail; supports device rotation; optimized for Retina display; (iOS 6.0 or later).

SHEET MUSIC NOTATION AND SHEET MUSIC READER APPS

A.P.S. MusicMaster Pro (http://apsdevelopmentllc.com/products/music masterpro/) Includes a PDF viewer and annotator, chromatic tuner with pitch pipe, audio recorder and player, metronome, timer/stopwatch, instrumental/ vocal ranges and transpositions, common guitar chords, fingering charts, musical terms and translations, PDF viewer with markup capabilities, and the ability to import PDFs from the Internet or other apps; (iOS 6.1 or later).

Avid Scorch (http://avid.force.com/pkb/articles/en_US/faq/en391291) App transforms iPad mobile device into an interactive music stand, score library, and sheet music store; generates interactive notation; give a performance using Music Stand mode; showcase own music; purchase scores from publishers; practice and perform favorite sheet music on the go; included content and how-to material; view own Sibelius or Sibelius First scores and lead sheets plus chart-toppers, classics, and other content from a variety of publishers; learn songs or adapt them to instrument by transposing a score, changing instruments, or converting to and from guitar tab; display individual parts or see the whole score on the Keyboard display; hear own score by playing it back with sampled instrument sounds; mix score; give a performance and turn pages in Music Stand mode; increase contrast and make music more legible by changing the text and paper texture; send files to Scorch from apps such as Dropbox or Mail; publish own music and sell it in the Scorch Store (requires Sibelius or Sibelius First); in-app purchases through iTunes; online FAQs and support forum; variety of paper textures and font choices; (iOS 5.0 or later).

DeepDish GigBook (http://deepdishdesigns.com/gigbook) Accessible mobile music library; keeps scores, songbooks, charts, and lyric sheets; add, sort, and organize; (iOS 5.1 or later).

forScore (http://forscore.co/) Sheet music reader app for practice and performance; create links to handle repeats with a single tap; play along to an audio track; use half-page turns to see the bottom half of the current page and the top half of the next one at the same time; can download any PDF file directly from the web or Dropbox; keep scores organized by tagging files; dynamic menus; create, share, and play through set lists for a manual approach to grouping and arranging files; create and edit drawing presets; add text to page with adjustable formatting; add common musical notation symbols; draw or import own designs; share files and annotations with friends via e-mail, Bluetooth, or AirDrop or print with AirPrint; (iOS 7.0 or later).

iComposer (http://soundplayground.com) Music notation tool for sheet music reading and editing; read, create, or share sheet music; more than eighty music elements such as notes, rest, tie, slur, dot, beam, triplet or tuplet, clefs, time signatures, key signatures, dynamics, lyrics, and many other symbols; multi-staves, multi-parts, and multi-voices for real music scores; export and import MusicXML files via e-mail or support website; play back with more than 130 instrument sounds; record humming and transcribe the melody to musical notations; easy and precise music input method for finger touching; (iOS 3.0 or later).

iWriteMusic (http://iWriteMusicApp.com/) Full-scale music notation editor; prepare sheet music for rehearsals, music lessons, home assignments, and more; practical and intuitive user interface; supports most major music notation elements; options to adjust page layout; creates parts by using hide track feature; easy transpose and partial transpose; printouts directly via AirPrint; exports PDF or JPG via e-mail; exports Standard MIDI file via e-mail; exports/imports iWriteMusic files via e-mail; multi-voice writing for up to four voices; track groupings; multi-bar rests; octave clefs; loop and swing options for playback; Touch+Hold to move note head; Quick Help; (iOS 7.0 or later).

Learn to Write Music (http://www.iphone-ipad-iapp.com/ipad.html) Draw with finger or stylus the main signs of musical notation; blank scores for class work; (iOS 7.0 or later).

Musical Notation iPhone Apps by Ambroise Charron Multimedia (http://www.iphone-ipad-iapp.com/iphone.html) *Neumes:* App that shows the neumes with their names, old notations, square notations, messine notation, beneventan notation, aquitanian notation, and modern equivalents; *Psaltis:* App that shows the neumes of Orthodox Byzantine music, including note names and definitions; *Mensural:* App that shows the mensural notation system including notes, ligatures, mensurals signs, keys, and symbols; *Maqam:* App that shows the maqam notation system (ajnas and maqam); *Raga:* App that shows the main ragas in Western musical notation; all apps are Retina display compatible; (iOS, varies by app).

Musicnotes (http://www.musicnotes.com/ipad/) Portable sheet music library; access to sheet music scans and PDFs as well as Musicnotes.com library of sheet music; built-in markup tools to draw, highlight, and type on sheet music; customizable set lists and folders management system; import own PDF sheet music collection with the "My Personal Library" add-on; link to Musicnotes.com account to import compatible sheet music files; high resolution; professional arrangements and engravings; transpose the key of Musicnotes sheet music; easy page turns by swiping finger to flip pages; jump to any page; zoom in on notation; options for different paper types; options for different page change methods including swipe or tap; supports the AirTurn BT-105 hands-free Bluetooth page-turner pedal; Conductor Mode: turn pages remotely and send markup notes to multiple iOS Devices over WiFi or Bluetooth; (iOS 4.3 or later).

MusicReader (http://www.musicreader.net/) Digital music stand app for the iPad; manage and use sheet music; PDF support; all settings and annotations are stored in the PDF; view PDF sheet music; library to catalog files; playlist support for rehearsals and performances; make annotations in the PDFs using multiple colors and line widths; standard symbols and support for redo/undo; display and turn pages in multiple ways; zoom into the sheet music; easy download of PDF files from websites; support for foot pedals to turn pages; use metronome, tuner, music player, and sound recorder; display sheet music on external screen; Dropbox synchronization; Bluetooth file transfer; (iOS 5.1 or later).

NotateMe (http://neuratron.com/notateme.html) Music composition and notation software featuring handwritten music recognition; quickly and accurately enter music notation with finger or stylus; simple, intuitive interface; similar to writing with pen and paper; instant playback and editing; score suitable for printing and publishing; for learning or teaching music notation; e-mail PDF, MusicXML, and MIDI files of scores to friends or other musicians; open in other apps and desktop notation packages; automatically sync scores to Dropbox to edit on phone; write music for solo instruments including voice and piano or ensembles from string quartets and choirs to full orchestral scores; includes support for transposing instruments; recognizes a wide range of music symbols, including notes with solid, open, and slanted noteheads, flags, beams, ledger lines, multiple voices per staff, chords, rests, accidentals including natural, sharp, double sharp, flat, articulation marks, augmentation dots, ties, slurs, hairpins, clef changes, and key signatures; barlines, clefs, time signatures, and tuplets are added automatically; add chord-symbols, dynamics, and lyrics to score; write comments or ideas in "red pen"; printed score updates automatically; (iOS 6.0 or later; Android 4.0.3 and up).

Notion (http://www.notionmusic.com/) Notation editor and playback tool for iPad; easy-to-use music creation tool; composition app to compose, edit, and playback scores using real audio samples performed by the London Symphony Orchestra recorded at Abbey Road Studios; user-friendly interface and simple interactive piano keyboard, fretboard, and drum pad; includes popular instruments such as guitar, bass, and more; use the score setup tool or open Notion and Progression iPad files as well as import MusicXML, Guitar Pro, and MIDI files; alter score and playback using articulations, expressions, and dynamics; full-featured multi-track mixer with effects; share work with anyone by sending a Notion file to view, hear, and edit; other options include sending MIDI or MusicXML files, or e-mailing a PDF file that can be viewed and printed; enter, edit, and playback notation, tab, or both; includes piano, keyboards, electric guitar, acoustic guitar, and electric bass; MIDI step-time entry with MIDI device; record real-time MIDI input into score; with Audition Mode can use the on-screen virtual instruments to hear sounds before entering them into score; clean and intuitive user interface; support for Retina display; interactive piano keyboard, twenty-four-fret guitar fretboard, and drum pad; quick and simple selection palette; distortion and reverb effects; full audio mixer; full range of orchestral functions and articulations including staccato, flutter tongues, trills, vibrato, and more; full range of guitar functions and articulations including bends, vibrato, slides, hammer on, pull off, mutes, whammy bar techniques, bass slap, harmonics, and more; drag-and-drop cursor option; switch instruments; transposition; insert text, rehearsal marks, or lyrics; chord symbols and diagrams; rhythm slashes and swing; continuous and page view; "Undo" and "Redo" functions; chord and melody modes; delete and erase capabilities; enter and edit title and composer information; file sharing; save as an audio file; import .notion, MIDI, MusicXML, or GuitarPro 3-5 files; e-mail .notion, PDF, MusicXML, .WAV, AAC, or MIDI files; search for tabs, MIDI, or MusicXML from within the app; import from Dropbox; add more instruments and gear; help files; (iOS 5.1 or later).

Old Music (http://www.ambroise-charron.com/en/ipad-apps.html) Old forms of musical notation; groups neumes, including their names, their old notations, their square notations, and their modern equivalents; Byzantine notation including note names and definitions; Mensural notation including notes, ligatures, mensurals signs, keys, and symbols; (iOS 5.0 or later).

OnSong (http://onsongapp.com) Manage large collections of chord charts and lyrics sheets for band or worship team; pull and reorder lists and flip from one song to the next; transpose chords; built for live performance musicians; import songs directly from online sources such as Dropbox or add own songs with built-in editor; type the song and surround chords in square brackets in line with lyrics; automatically detects sections and titles; import existing songs using iTunes File Sharing or Dropbox; supports PDF, Word, Pages, JPG, PNG, TIFF, ChordPro, and text file formats; listen to songs playing directly from iTunes music library; play metronome or get music from iTunes with one tap; simple song and chord entry; import, export, and synchronize with Dropbox and others; pull and change songs for set lists; flip or tap through all the songs in a set; transpose and capo with the brush of a finger; highlight or bold chords for visibility; change font size and style; keep track of past sets and archive; import songs from online sources; share wirelessly over Bluetooth; send list via e-mail; print set wirelessly to AirPrint printer; works with many foot pedals such as the AirTurn BT-105, Griffin Stompbox, and iRig

Blueboard; project lyrics or use a stage monitor with VGA, HDMI, or AirPlay video support; (iOS 4.3 or later).

PDF Sheet Music Reader (http://www.rekono.com/pdf-sheet-music-reader/) Keep sheet music in a single place; no restrictions on which songs can be viewed and no additional purchases to make; view any PDF file by downloading it using the built-in web browser or importing the file using iTunes; preloaded with songs; add own PDF files; edit song titles and composer names; delete PDFs from library; tap or swipe to flip pages; audible time signature metronome; draw on sheet music to record trouble spots; organize songs into set lists; (iOS 3.2 or later).

Reflow Score Writer (http://www.reflowapp.com/) Tablature and sheet music editor to create original scores; playback using embedded software synthesizer; import and play any Guitar Pro, PowerTab, or MIDI file; modify song while playing; compose and practice any song for guitar, bass, drums, piano, banjo, and many other instruments using standard notation or tablature; iCloud and Dropbox support; Retina enabled; export to MIDI, GP5, PDF, and share files by e-mail; write music using tablature, standard notation, chord diagrams, and chord names; support of piano grand staff; musical directions including coda, segno, etc.; use different scores for each instrument of a song; full vector graphics PDF export for manipulation in Illustrator; guitar effects and articulations including bend, brush, slides, and more; (iOS 6.1 or later).

Score Creator (http://simplesongcreator.com/) Music composing application designed for mobile platform; interface allows user to input notes and chords quickly; virtual note keyboard and chord board; write music notation; supports both treble and bass clef with a wide range of notes and music symbols including note duration, time signature, key signature, slurs, ties, etc.; write lyrics and chord symbols; multiple tracks with different instruments including piano, guitar, violin, and saxophone; score for transposing instruments including saxophone (soprano, alto, tenor, baritone), B♭ clarinet, and B♭ trumpet; playback sound for each instrument; transpose songs into any key; import songs from MIDI files; export songs to MIDI or MusicXML files; copy exported files to computer through iTunes File Sharing and continue editing the songs using professional music composing applications like Finale, Sibelius, Encore, and MuseScore; send files via e-mail; send exported MIDI files to other apps; open MIDI files from within other apps; export to PDF; editing assistant features including multiple selecting notes, copy and paste, undo and redo; iCloud storage; (iOS 6.0 or later).

Scorio Music Notator (http://www.scorio.com/web/scorio/) Write down melodies, lead sheets with chords and lyrics, arrangements, and full scores;

enter and edit music intuitively with fingertips; short touches select notation elements; long touches let user insert new notes into melodies and chords; scrolling, zooming, page turning, and orientation change; edit lyrics and chords and modify the score structure with the dialog area on the right side of the app; works with Scorio notation portal on the web and requires an Internet connection; write and save a score within the app and find it online in Scorio account; after saving, can edit compositions in any web browser and outside the app on PC or Mac; insert and edit notes and other notation elements; virtual keyboard for entering notes; insert and edit chord and fret symbols; display and edit lyrics; export scores as high-quality PDF files; select from nineteen score templates; edit score structure by adding and/or deleting staves; transpose scores; automatic part extraction; MIDI playback with 128 selectable MIDI instruments; load scores from the Scorio database; publish scores; work on scores in account on PC, Mac, and other tablet devices; (iOS 5.1 or later; Android 4.0 and up).

Sheet Music Direct (http://www.sheetmusicdirect.us/ipad/) Sheet music and guitar tab viewer and score library with access to more than one hundred thousand high-quality scores; includes fifteen songs for free; import scanned sheet music PDFs from personal library via iTunes, e-mail, and Dropbox; score playback; multi-track mixing; key transposition; transposable melody line; one-touch page navigation; set lists; social media sharing tools; metronome and tuner; create a Sheet Music Direct account from within the app to view and sync all purchased sheet music from iPad and computer; in-app purchases will automatically be backed up; view, sort, and search all sheet music; tapping on a song will open it in the music viewer; two library viewing modes include List View and Browse View; three sorting options by title, artist, or notation; move songs into archive to not view in active library; create, view, and edit set lists; customize Music Viewer paper type and background; manage social media credentials and push notifications; verify app version; hands-free page turning using Bluetooth pedals; supports the PageFlip Cicada and the AirTurn BT-105; access more than one hundred thousand songs arranged for a variety of instruments including piano, voice, guitar, bass, and ukulele in a variety of styles including pop, Broadway, classical, jazz, rock, country, and more; search, browse, and preview the first page of the score in the Music Viewer; (iOS 4.3 or later).

Sheet Music Plus (http://www.sheetmusicplus.com/help/iPad) Access previously purchased digital print titles on iPad; free Sheet Music Plus Digital Print Viewer; sort titles by purchase date, title, or artist; turn the page with a finger swipe or the press of a button; skip to the last page or jump back to the first page; jump to any page of music by page number; zoom in and zoom out;

adjust view to landscape or portrait; new purchases appear automatically in library; carry entire digital print sheet music library to rehearsals, jam sessions, and music lessons; preserve music from water damage, stains, smells, and natural wear; replace damaged or forgotten music and bulky music folders; library of more than 140,000 digital print titles in twenty-two different styles of music including new pop releases, classical favorites, and more; (iOS 4.0 or later).

SheetRack (http://www.sheetrack.com/) Sheet music reading app; library with access to hundreds of music sheets for free; import own scores; sheet music reading and score-on-the-go experience; import PDF sheet music and/ or download free sheet music from library; organized view of all sheet music downloaded with a book-like image index; audio and visual metronome; search by sheet title, composer, or genre; sheet statistics including number of pages and number of times read; turn sheet pages by blowing on the microphone hands-free; ad free; imported titles are marked to distinguish them from downloaded ones; large library of public domain scores including solos, duets, orchestral arrangements, etc.; previews for each title; composer portraits and genre symbols list; rating and number of pages of each sheet displayed along with the title for reference; rate sheets individually; search engine to search for scores locally and online, by title, composer, or genre; small menu to access all features in one place and display information on the sheet music being displayed; pages of sheets downloaded are automatically tilted and formatted to the iPad screen resolution; remembers the last page playing; pan and zoom on sheet pages; listen to external music while the app is active; downloaded or imported sheets are saved locally and permanently until deleted by the user; efficient in dealing with multiple music sheets; Universal app; (iOS 3.1.3 or later).

Sibelius Essentials 1 (http://www.makemoremusic.co.uk) Covers the basics of working with Sibelius including ten tutorials on creating a score, navigating, selecting and editing, various methods of note entry, recording into Sibelius with Flexi-time, playback controls, marking up the score, adding dynamics and articulation, preparing arts, layout, and formatting; (iOS 3.0 or later).

SongBook for Finale (http://www.finalemusic.com/products/finale-song book/) Bring entire music collection to every practice, rehearsal, and performance; hear sheet music play back; print copies and parts; transforms iPad into an interactive music folder to view, play, and print music scores created by Finale products; access all parts from within score files; create set lists for quick access to music during performances; hear music play back with the nuance of human performers; control tempo; navigate via pinch-to-zoom, double-tap, and swipe-away gestures; share files via e-mail; flip pages hands-free via AirTurn page turners; (iOS 5.0 or later).

Tonara Sheet Music (http://tonara.com) Interactive scores that listen to the user play, mark the position in the score, and turn the pages automatically; hundreds of free classical scores and songs; access to thousands of classical, rock, pop, soundtracks, and other scores in the Tonara store; music synchronization capabilities; recordings automatically synchronized to score to review practice; listen to MIDI sample playback synchronized to the score; gather all scores in one place; upload PDF files from different sources and have all sheet music organized in one place; import entire music collection from Dropbox, Google Drive, e-mail, or web browser; edit the information; flip the pages using an AirTurn or PageFlip external pedal; visual graphical reports of progress; upload or import scores from private collection; digital sheet music binder; listen back to sessions; share performances; resilient to background noise; tempo display; visual and audio metronome; large selection of sheet music for free and for purchase; scores formatted for the iPad; for musicians of all levels; Retina display; (iOS 7.0 or later).

Virtual Sheet Music (http://www.virtualsheetmusic.com) Download digital sheet music and play offline as well as listen to and download thousands of MP3 audio files; available for most music for two or more instruments; page turning system allows app to be used during live performances; more than fifty free sheet music items included; online complete catalog of high-quality digital sheet music to download and store titles on device including PDF, MP3 files, and MP3 accompaniment files; search the catalog by keyword; organize music in folders; display detailed information about any music item; more than forty thousand MP3 files; send music via e-mail as a PDF file or print out via AirPrint; import external PDF files via iTunes sharing feature; access music through the log-in feature and synchronize downloads; single-touch page turning system compatible with AirTurn Bluetooth foot pedal devices for live performance; download high-quality digital, traditional, and classical sheet music with MIDI, MP3 audio, and MP3 music accompaniment files; requires Internet connectivity to access the online catalog; (iOS 6.0 or later).

· 2 ·

Composing, Arranging,
and Improvising Music Apps

COMPOSING AND ARRANGING MUSIC APPS

Arpeggionome (http://alexander-randon.com/arpeggionome-for-iphone/) Design an arpeggio pattern with various knobs and buttons; move up/down the matrix to set the speed or left/right to set the pitch; tilt to control volume and pitchbend; shake for vibrato; use to control other iOS music apps, external gear, and computer software; Audiobus supported input; (iOS 5.0 or later).

Bass Line Composer (http://drumlineapps.com) App includes a bass drum line to play along with; practice splits and interp; write out exercises or make up new ones; speed up, slow down, try it out, edit it; compose and play along with the actual bass drum sounds of Paul Rennick as recorded by bass tech Brian Lowe; best used with headphones; fully featured tool with time signatures, variable splits, and more; sound palette including flams, shots, muffle, buzz, and rim clicks; range of dynamics from *p* to *ff*; (iOS 4.3 or later).

Bloom HD (http://www.generativemusic.com/bloom.html) Part instrument, part composition, and part artwork; innovative controls allow user to create elaborate patterns and unique melodies by tapping the screen; generative music player takes over when app is left idle, creating an infinite selection of compositions and their accompanying visualizations; preserves the sounds and features of the original version, but adds a new interface that allows the users to immerse themselves in the full-screen experience or to continue playing on a smaller area while adjusting the controls; Classic, Infinite, and Freestyle modes; twelve different mood settings; three sounds; random mood shuffle; adjustable delay; evolve when idle; Sleep Timer; recommended use with headphones or external speakers; (iOS 4.3 or later).

Componendo Music Arranger (http://www.componendo.eu/project/home .html) Music app in arranger style; includes piano and sax to play melodies and keyboard and bass for the accompaniment; twenty-five drum loops for the rhythm; accompaniment chords are laid out on a grid, not a sequencer; by pressing a few buttons can create an accompaniment; will recommend chords to use while making up turnarounds; designed to play live; (iOS 4.3 or later).

Composer (http://www.ifomia.com/apps/composer/index.html) Compose music and e-mail the music notation sheet and the generated sound file to anyone; edit the music score using fingers; change key signature, tempo, meter, and edit the title and other information for the music score sheet; in full version, user can have multiple tracks for one song with different meters, tempo, or key signatures and add lyrics; rhythms, chords, and tuplets are also supported; save the current track and edit at a later time; (iOS 5.0 or later).

ComposerBase (http://www.composerbase.com) View, print, e-mail, and record music scores; showcase compositions; buy music collections to build up library; easy page turns with one swipe to flip pages; record, save, and share audio files; page slider to jump to any page; files are backed up in the Cloud; record practice sessions; music collections available: Albéniz, Bach, Beethoven, Chopin, Czerny, Debussy, Grieg, Joplin, Liszt, Mendelssohn, Mozart, Rachmaninoff, Ravel, Schubert, Schumann, Scriabin, Tchaikovsky, and Verdi; (iOS 5.1 or later).

Dingsaller (http://timbolstad.com/dingsaller/) Workspace for creating algorithmic compositions; patch together a complex graph of music generating nodes, synths, and effects; synths and sampler instruments; digital effects; pattern sequencer; random note generators; touch and tilt controllers; MIDI input and output; recording; audio copy/paste; (iOS 5.0 or later).

Gestrument (http://www.gestrument.com) Improvise or compose within scales and rhythms; use parameters like pulse density, scale morphing, rhythm randomness, or pitch fluctuation; use the tutorials or define own settings to fit musical style; play up to eight instruments at once all with different individual settings; supports AudioCopy and General Pasteboard; Audiobus supported input; Apple Core MIDI compliant; MIDI Clock Sync; Virtual MIDI port; (iOS 5.1 or later).

GlowTunes (http://www.apphappystudios.com/apps) Combines light and sound to create works of art; select the instrument and draw anything by turning on and off light pixels; press Play and watch as drawing comes to life and creates a melody; twelve instruments to choose from; four included with initial purchase including piano and music box; turn on Rainbow Mode to draw with rainbow patterns; play speed controls; educational and entertainment value; HD graphics for Retina display; Universal app; (iOS 4.3 or later; Android 2.2 and up).

HyperTunes (http://hypertunes.net) Modular MIDI arranging for iOS; collect and organize musical patterns, song fragments, and songs within the app; compose and arrange music in terms of musical structure, such as verse-chorus-bridge song form; import MIDI files via iTunes File Sharing; record MIDI from other apps via Virtual MIDI or from other devices via Network MIDI; project browsing and auditioning; browse to any instrument part or song section in any project; pair wise auditioning; hear how any two parts sound together or in sequence; playlist feature; audition parts, part combinations, or sections playlist style; play back one song with the structure of a different song; edit song structure; insert, move, or remove entire sections in one step; set song form; pick a form from the menu and apply to any song in one step; split songs into sections; break up an imported song into sections; one-touch play-back of entire song, individual sections, or any instrument part of a section; export MIDI files via iTunes File Sharing or e-mail, then import into desktop studio; includes a set of basic patterns for drums, bass, and guitar; import own MIDI files or third-party MIDI libraries; (iOS 5.0 or later).

Konkreet Performer (http://konkreetlabs.com/overview/) App uses a shape that is a visual representation of the sound; user can sculpt using multi-touch and gestures; new method of musical expression; Virtual MIDI port; Apple Core MIDI compliant; (iOS 5.1 or later).

Maestro Touch Free (http://www.maestrotouch.com) Digital instrument designed for touch-screen interfaces; designed for novice, intermediate, and expert musicians; create masterpieces regardless of musical background; interface maps directly to many of the underlying concepts of music theory; (iOS 5.0 or later).

Melody Composer (http://patrickqkelly.com/index.php/ipad/melody-composer) Write a melody in standard music notation; write in treble, alto, tenor, or bass clef; change a melody's clef; touch and play along on an A♭, E♭, B♭, F, C, G, D, A, or E instrument; (iOS 7.0 or later).

Melody Music Maker (http://www.melodymusicmaker.com) Transforms a simple melody into a well-composed and correctly arranged piece; start with an idea for a melody; remaining part will be handled by the app; arranger and composer in one; (iOS 7.0 or later).

MorphWiz (http://www.wizdommusic.com/products/morphwiz.html) Musical instrument and visual experience; created by Jordan Rudess and Kevin Chartier; contains more than fifty presets; synthesis, effects, and visual options allow the instrument to be configured to the user's tastes; Audiobus supported input; supports AudioCopy and General Pasteboard; (iOS 4.3 or later).

Music Box Composer (http://www.jellybiscuits.com/?page_id=705) Music composing tool based on the physical fifteen-, twenty-, and thirty-note paper strip music boxes; allows user to compose, preview, play, save, and share creations; (iOS 4.3 or later).

Musyc (http://www.fingerlab.net/website/Fingerlab/Musyc.html) Touch turns into music; draw shapes and listen to music while viewing sounds bouncing on the screen; includes eighty-eight instruments organized into twenty-two groups; Retina display; audio track mixer with level, pitch, length, pan, and mute; two effects channels with ten effects including Delay, Overdrive, Reverb, and Phaser; physical sequencer; motion recorder; user sound kit with samples; import from Dropbox, microphone, iPod, iTunes, or AudioPaste;

real-time audio recording; high-quality or compressed exports including Dropbox, SoundCloud, Mail, AudioCopy, and iTunes; video recording and export to Facebook, Vimeo, Mail, and Photo Library; Inter-App Audio support; Audio background support; two skins available; (iOS 7.0 or later).

Noatikl 2—Generative Music Lab (http://intermorphic.com/noatikl/) Interactive composition tool; templates included; Audiobus supported input; (iOS 7.0 or later).

NodeBeat HD (http://nodebeat.com) Visual music app; create music or listen to app generate its own; Virtual MIDI port; save and share creations with others; supports AudioCopy and General Pasteboard; Audiobus supported input; Apple Core MIDI compliant; (iOS 6.0 or later).

Orbita for iOS (http://keijiro.github.io/orbita/) Music-making app creates sounds by tapping the screen of a mobile device; simple and intuitive; variety of possibilities; (iOS 4.3 or later).

Orphion (http://www.orphion.de) Musical instrument with a unique sound between string instruments and percussion; can create expressive sounds and melodies by moving fingers on virtual pads; control other apps via Virtual MIDI; supports AudioCopy and General Pasteboard; Audiobus supported input; Apple Core MIDI compliant; Virtual MIDI port; (iOS 7.0 or later).

PatternMusic MXXIV (http://www.patternmusic.com/frontdoor/Pattern Music.html) Three music tools in one; create songs intuitively using layered instrument patterns; play an interactive polyphonic musical instrument with more than fifty different voices; jam along with iPod music library; The Stage: mix and arrange songs by moving instrument icons; add, duplicate, and delete instruments; Multi-Touch Pattern Editor: create unique looping note patterns; editor zoom; up to sixty-four notes per pattern via the scrolling timeline; scroll through a four-octave range for each instrument; Solo Mode: focus on a single part of a composition; dynamics for individual notes; change voices; full drum kit and percussion set; built-in help and tutorials; Instrument and Song Settings; tap tempo for interactive tempo matching; time signature; more than forty scales; note length configurable from sixteenth-note triplets to dotted whole notes; infinite or counted looping control; configurable instrument start point; shuffle effects; transpose up or down two octaves; eight-octave range; song management including song duplicate, rename, and delete; auto-save; (iOS 3.2 or later).

PianoStudio (http://frontierdesign.com/PianoStudio/AppStore) Explore musical ideas, learn songs, write tunes, and play along with iTunes music library; place individual notes, chords, and musical phrases onto a set of buttons to arrange and play; write and perform piano music with chord progressions, two-handed patterns, and melodies; supports AudioCopy and General Pasteboard; (iOS 3.1 or later).

Snare Line Composer (http://drumlineapps.com) Input and play along; change tempos, edit, revise, save, listen, learn; tweak licks; add rimshots, buzzes, and dynamics; (iOS 4.3 or later).

SoundMakers (http://www.soundmakers.ca/soundmakers-ipad-app) Allows user to create and share new two-minute compositions based on samples from archive of commissioned works; mix and manipulate a large range of samples; (iOS 6.0 or later).

SoundPrism Pro (http://www.soundprism.com) Composition tool and MIDI controller; interactive generator for harmonic sound textures; control existing VSTs, synthesizers, and sound modules; visualize and internalize musical relationships; interact musically with geometric pitch space visualizations; define pitch layouts with the PitchClassShifter; Audiobus supported input slot; Apple Core MIDI compliant; Virtual MIDI port; (iOS 4.3 or later).

Soundrop (http://maxweisel.com) Musical geometry; draw lines and watch as app uses them to create music; listen with headphones for the full effect; tap and drag to draw a line; drag the handles to edit or double-tap to delete; drag dropper to reposition drop point; audio pans left/right with bounce location onscreen; (iOS 3.0 or later).

SpaceWiz (http://www.wizdommusic.com/products/spacewiz.html) Interact with and control an environment at the highest level of complexity; with the manipulation of fingers on the playing surface and the technology of Chaos and Satellite generators, user fabricates particles that collide into planets; includes an onboard real-time synthesizer with an independent set of sounds and parameters; supports AudioCopy and General Pasteboard; (iOS 5.1 or later).

studio.HD (http://www.soundtrends.com/apps/studiohd-1) Multi-track, multi-touch music studio for songwriters, producers, and musicians; twenty-four stereo tracks of playback with recording, editing, loops, automated mixing, and effects; record rap, vocal, or guitar; overdub and produce more tracks; effects and mixing; supports AudioCopy and General Pasteboard; (iOS 4.2 or later).

StudioApp Pro (http://www.krasidy.com/studioapppro.php) App combines the functionalities of several Krasidy apps to make beats, record vocals, remix, and more; (iOS 4.0 or later).

StudioMini XL (http://www.studiomini.me/StudioMini/StudioMini___ Info___Recording_Studio_App_for_iPhone_iPad_iPod_touch.html) App is a recording studio with CD-quality audio, pro audio loops, a metronome, and more; (iOS 3.2 or later).

Symphonix Evolution (http://www.vinclaro.com.au/ios/symphonix/symphonix.htm) Create music by dragging notes directly onto the score; record or play back via MIDI devices or using the inbuilt virtual synthesizer; Virtual MIDI port; Apple Core MIDI compliant; (iOS 6.0 or later).

Symphonix Evolution Player (http://www.vinclaro.com.au/ios/symphonix_player/default.html) Basic MIDI file player that supports General MIDI files using the GM/GS specification with up to sixteen tracks; play Symphonix Evolution files created in app using a suite of more than 150 wavetable instruments; Apple Core MIDI compliant; (iOS 5.1 or later).

Symphony Pro (http://www.symphonypro.net/) Pioneer in touch-based music composition; compose complex orchestrations, sheet music, lead sheets, chord charts, and guitar tabs on the iPad and hear them played back; record a part in real time with the built-in keyboard, guitar, or bass guitar to capture a musical idea and have it automatically transcribed into sheet music; import a MusicXML, MXL, MIDI, or Symphony file from computer or from another app; create a new project based on a template or create a brand new project; to add a note, select a note length and tap anywhere on the score; use the built-in instruments or a Core MIDI compatible keyboard to enter notes and chords in normal editing mode or in real-time recording mode; add notational elements to compositions including different dynamics and articulations, time and key signature changes, clef changes, alternate endings, repeats, double accidentals, double and triple dotted notes, arpeggiated chords, glissandos, tuplets, tempo changes, coda markings, pedal markings, alternate noteheads, similes, and pickup measures; create and edit chord symbols, lyrics, and annotations using the iPad's standard keyboard; print out score or export it as a PDF, MusicXML, AAC, MIDI, or Symphony file; share files via e-mail or directly to a computer with iTunes File Sharing; save images of score directly to the Photos app and share them through Photo Stream; 114 built-in instruments; up to fifteen instruments, forty-five staves, and four voices/layers per composition; print the entire score from iPad with AirPrint; edit, view, and play score in portrait or landscape mode; edit multiple notes and rests at once; multi-measure, multi-track copy/paste functionality; delete a note, measure, or an entire selection area at once; familiar gestures for editing and navigating; undo and redo support; fully customizable stem directions and note beaming; transposing instruments and concert pitch options; change the tempo or clef at any point in the score; stream audio or share screen wirelessly with AirPlay to compatible speakers, stereo systems, AV receivers, Apple TV, or other devices; manage documents in a bookshelf display or in a list with a built-in index; create custom project template or use a built-in one; customizable track groupings for SATB and instrument families; play a built-in instrument without entering any notes; customizable chord symbol, lyric, and annotation sizes; loop playback between specific measures; metronome click during playback and recording with a customizable volume level; auto-saving; built-in manual; (iOS 5.1 or later).

Tenor Line Composer (http://drumlineapps.com) Sweeps, exercises, street beats, and solos; work out ensemble together; slow down, speed up, edit; composer playground for ideas; play back with the actual tenor drum sounds of Paul Rennick's 2012 Santa Clara Vanguard, as recorded by bass tech Brian Lowe; best with headphones; (iOS 4.3 or later).

WI Orchestra (http://www.wallanderinstruments.com/?mode=apps&lang=en) Full orchestral library and composition tool; create and record orchestral music using the instruments and built-in recording features; choose between a full range of orchestral strings, brasswinds, woodwinds, saxophones, and percussion; velocity-sensitive keyboard with dynamics presets; export songs to computer; each instrument comes with a picture and description of the instrument; use the microphone as a breath controller; Universal app; (iOS 3.0 or later).

IMPROVISING MELODIES, VARIATIONS, AND ACCOMPANIMENTS APPS

Alfred Play Along (www.jammit.com) Play favorite songs on concert band instrument; rebalance the mix to hear every detail of the performance; true instrument isolation; scrolling notation: audio stays in sync with scrolling score; advanced tools: variable slow-down mode and section looping to focus on difficult musical passages; recording; (iOS 5.0 or later).

Band (http://moocowmusic.com/OtherApps/index.html) Collection of virtual instruments that lets user play and record tunes; includes piano, bass guitar, and drums; "Twelve Bar Blues" instrument; play musical scale by sliding finger up and down the screen; (iOS 3.0 or later).

BandMaster HD (http://ipad.rogame.com/pages/BandMaster.html) Like having a whole band; includes bass, drums, piano, samples, and several styles for each instrument; create songs with the available templates; songs are displayed in song list where user can launch playback or edit views; playback view displays song's chords in a grid that scrolls down and highlights each measure as it is played; user determines the speed and number of choruses as well as the style; can turn off any of the instruments; use band's logo as the background for the playback view; includes exercises and blues and jazz standard forms of different lengths; can add and delete measures, transpose the song, or switch to a different time signature; practice soloing; intuitive interface; supports all common chords; American and European chord notation; global transposition for horn players; more than twenty-six song templates; song exchange with Mac OS versions of BandMaster; (iOS 6.0 or later).

Changeling Sequencer (http://changelingapp.wordpress.com) Diatonic chord sequencer; helps by providing the basic tools to improvise; improvisation is a skill that can normally only be obtained by investing a lot of discipline and time into studying music theory and practice; app does not generate compositions for the user; all recorded musical notes remain relative and can be changed in various ways during playback; basic instructions: start by recording a simple melody line with one finger; change the scale and root by sliding the ribbons; transform the melody to a chord progression by turning the chord knob; play around with the available voicings and transition modes that affect chord inversions; spread out the chord notes over time using the arpeggiator; switch to the next track by swiping down on the pattern area; change to playback mode and start improvising a solo over the chord progression; sixteen MIDI channels to record patterns to DAW or other MIDI capable iOS apps; MIDI Clock Sync; Virtual MIDI port; (iOS 5.1 or later).

Chops (http://tonalapps.com/chops/) Improve musical improvisation; practice aid for musicians; enter the chord changes to play over or select chord changes from the library of standard chord progressions; app will "strike up the band" in any of sixteen jazz and rock styles in all twelve keys and at any tempo; designed to be used with speakers or earphones; (iOS 4.0 or later).

Chordbot and ***Chordbot Pro*** (http://chordbot.com) Songwriting tool and electronic accompanist that lets user create and play complex chord progressions; add chords, select a comping style, and hit play; for songwriting experiments or to create backing tracks for practicing solos; create arrangements and mixes by selecting from hundreds of piano, guitar, synth, bass, and drum tracks; more than sixty chord types in all inversions; more than seventy comping style presets; simple and usable arrangements; more than four hundred instrument tracks; individually mixable dynamic instrument tracks that can be combined to form unique comping styles; inversion control slash chords; manually select chord inversions or let app select the inversion automatically; songs can be structured into repeatable sections including verse and chorus; each section can have its own comping style and arrangement; MIDI/WAV export and background playback in a DAW/MIDI sequencer or play along with a virtual instrument; Song-O-Matic song generator; see website for more information and audio demos; (iOS 4.3 or later; Android 2.2 and up).

CutieMelody (http://cutieapps.com/cutiemelody/) Simple touch-screen version of a children's glockenspiel; realistic sounds and responsive multi-touch playing and sliding; improvise freely or perform favorite melodies; animated bars and note letters provide feedback; floating notes in the background; for kids to learn the basics of making music; (iOS 3.2 or later).

iImprov—Bebop (http://jazzappsmobile.com/?page_id=40) Explores the language of jazz as it applies to playing bebop; lessons cover the use of placing half steps to create longer lines as well as breakdowns of rhythm changes, the "Confirmation" sequence, turnarounds, and basic substitutions; realize the bebop scale through strategic placement of half steps to include all twelve notes of the chromatic scale while maintaining appropriate chord colors; nine comprehensive lessons and eight JAM-A-longs; each lesson contains a detailed explanation as well as exercises, transposable notation, and tempo-adjustable audio examples; concepts demoed by saxophonist Tony Bray; (iOS 4.0 or later; Android 2.1 and up).

iImprov—Chord/Scale Compendium (http://jazzappsmobile.com/?page_id=46) Reference app that provides a visual and audio representation of the most useful scale to chord choices for jazz improvisation; ear training tool for learning the different scale and chord colors; organizes twenty-five different chord qualities into three families: Major, Minor, and Dominant; when a specific chord quality is selected, that chord is displayed in standard notation along with a key selection wheel; depending on the chord, from one to five scales are recommended; fast forward button cycles through the recommendations; play button plays the currently selected chord and scale; chords are displayed in close position (stacked thirds), but the chord is sounded with a more practical voicing that might be encountered in a playing situation; in addition to written scales for each chord type and key, the scale can be played over a typical voicing for each chord; bass or treble clef and the appropriate transposition for the instrument can be selected; all scales and chords may be presented in any key; (iOS 3.2 or later; Android 2.1. and up).

iImprov—Contemporary Colors (http://jazzappsmobile.com/?page_id=42) Focuses on handling the chord changes that the user will find in the jazz repertoire from post-bebop to the present; gives user the tools to take on chord progressions found in tunes by Wayne Shorter, Chick Corea, Herbie Hancock, and any other composer that has ventured away from the writing of chord changes that would be typically found in the standards; these types of chord progressions would include highly altered chords, slash chords, and chords created from scale colors outside of diatonic major; lessons cover in detail how to use diminished, whole tone, augmented, Lydian augmented, and harmonic minor scales to play over these types of progressions; includes exercises, patterns, demos, and JAM-A-longs with balance control; transposable notation and tempo adjustable audio; many of the concepts are demoed by saxophonist Tony Bray; JAM-A-longs include individually created files for each key that will loop indefinitely; all keys are covered; tempos are adjustable; can set the overall transposition for a specific instrument; (iOS 4.0 or later; Android 2.1 and up).

iImprov—Modal (http://jazzappsmobile.com/?page_id=38) Contains a full explanation of concepts and theory as well as written and audio examples and exercises; select the correct transposition for the instrument; provides the tools for playing over modal tunes; exercises and JAM-A-longs specific to each type of tune are included in all keys with mixer and tempo control and unlimited loop length; detailed theory of note to chord relations and use of modal scales; detailed theory of how to use pentatonic scales; audio and visual representations of the related scales; audio playback of scales played over chords; exercises that include live audio examples from saxophonist Tony Bray; (iOS 3.2 or later; Android 2.1 and up).

iImprov—The Fundamentals (http://jazzappsmobile.com/?page_id=33) Helps user learn the language of jazz improvisation; covers the basics of the jazz jam session, basic symbology, creating scales, diatonicism, form, and scale to chord relations; basic ear training module; demos; basic tools to help build a foundation to further develop craft as an improvising musician; detailed introduction to basic theory; interactive lessons that include exercises as well as written and audio examples; JAM-A-longs in all keys with tempo and balance control; scale to chord compendium; supplemental material available on website including PDF files; (iOS 4.0 or later; Android 2.1 and up).

iImprov—The Minor II V (http://jazzappsmobile.com/?page_id=36) Focuses on the Minor II V chord progression, one of the more common progressions found in the jazz repertoire; practice method for playing a particular scale over the chords; JAM-A-longs include focus on individual chords as well as the whole progression; defines the scale to chord relationships for each key; allows the user to listen to the isolated scale as well as hear it over the appropriate chord; (iOS 3.2 or later; Android 2.1 and up).

ImproVox (http://www.museami.com/improvox/) Vocal instrument; real-time pitch correction and harmonization enable user to create harmonies as they sing and always sound in tune; add textures with reverb and echo effects; record performance and share with friends; state-of-the-art audio processing; real-time pitch correction; real-time harmonization; effects including reverb, echo, ring modulator, flanger, and more; recording and overdubbing; share via SoundCloud or e-mail; import tracks from iTunes Library; AudioCopy and AudioPaste support; Audiobus support; headphones required; (iOS 6.0 or later).

iReal b aka iReal Pro Music Book & Play Along (http://irealpro .com/#features) Tool to help musicians of all levels; simulates a real-sounding band that can accompany users as they practice; create, edit, print, share, and collect chord charts of favorite songs for reference while practicing or performing; practice with a realistic-sounding piano or guitar; bass and drum accompaniments for any downloaded or user-created chord chart; virtual band accompanies as user practices; more than thirty different accompaniment styles; accompaniment can include guitar, piano, bass, drums, vibraphone, and more; play, edit, and import songs; thousands of songs can be imported from the forums in a few simple steps; edit existing songs or create own with the editor; player will play any song that user edits or creates; create multiple editable playlists; practice at any level; includes fifty exercises for practicing common chord progressions; transpose any chart to any key or to number notation; loop a selection of measures of a chart for focused practicing; advanced practice settings include automatic tempo increase and automatic key transposition; Global E♭, B♭, F, and G transposition for horn players; share, print, and export individual charts or whole playlists with other iReal Pro users via e-mail and the forums; support for AirPrint; support for Audiobus; export charts as JPEG, PDF, and MusicXML; export audio as WAV; Audio Copy/Paste supported; AAC and MIDI; support for iRig, AmpliTube, and other guitar connection kits; support for AirTurn and other Bluetooth page turners; see website for demos; Jazz Styles Pack includes fifteen jazz play-along styles: Ballad Swing, Ballad Even, Slow Swing, Medium Swing, Medium Up Swing, Up Tempo Swing, Double Time Swing, Swing Two/Four, Bossa Nova, Even 8ths, Latin, Latin/Swing, Afro 12/8, Gypsy Jazz, Practice; Latin Styles Pack includes eight Latin play-along styles: Brazil-Bossa Electric, Brazil-Bossa Acoustic, Brazil-Samba, Cuba-Son Montuno 2-3, Cuba-Son Montuno 3-2, Cuba-Cha Cha Cha, Cuba-Bolero; Argentina-Tango; Pop Styles Pack includes twelve pop play-along styles: Rock, Slow Rock, 12/8 Rock, Funk, Soul, RnB, Smooth, Disco, Reggae, Shuffle, Country, Bluegrass; Guitar Chords: displays guitar chord diagrams during playback following the chord progression; extensive chord diagrams library; see songs with all the guitar chord diagrams in line; Piano Chords: displays piano chord diagrams during playback following the chord progression; extensive chord diagrams library; see songs with all the piano chord diagrams in line; Chord Scales: displays chord and scale diagrams during playback following the chord progression; extensive library of chord and scale diagrams; backup songs; (iOS 5.1 or later).

Jam Player—Time and Pitch Audio Player (http://www.positivegrid.com) App for instrument players, singers, dancers, and musicians; makes learning, transcription, and practicing songs easier by slowing down the tempo, adjusting the pitch, repeating loops; load a song directly from the iTunes library; speed control knob can be used to slow down the tempo of the song without affecting its pitch; pitch control raises and lowers the pitch but has no impact on the tempo; change the key of the original track to match the singer's vocal range; Audiobus supported input; (iOS 5.0 or later).

Jammit (http://www.jammit.com) Jam with favorite bands and songs; isolate or remove an instrument from the mix and jam with the original multi-tracks; (iOS 6.1 or later).

Jamn and JamnPRO—The Musician's Multi-Tool (http://getjamn.com/) Large library of music theory for guitar, bass, piano, and ukulele; patented design "Jamn Wheel" lets user transpose key and visualize music; two modules: eHands and the Multi-Tool; eHands is a virtual chord finder; open up and tap on the list to see virtual hands overlaid on the guitar; can turn the hands on or off to get a closer look at the chord fingering; the Multi-Tool is more advanced with high-tech features; swipe the instrument above the Jamn Wheel to learn more; wheel is a visual aid and key transposer; tool to help user learn and remember notes, chords, and scales of any key; plays real-life audio for every chord, note, and scale; Chords Library with more than 1,680 popular chords; built-in tuner works for guitar, bass, and ukulele; PROScales menu; (iOS 5.0 or later).

JamPad Plus (http://www.h2indie.com/) Digital jamming machine for the iPad; play up to four tracks at the same time without needing a live band; keyboard with seven octaves; full-length guitar fretboard with scale assistant; select a chord and pattern with autoplay rhythm guitar machine; drum beat maker comes with many preset beats or customize own; onboard mixer for mixing the tracks; tempo slider for adjusting the tempo of the rhythm guitar and drum beat in real time; (iOS 5.0 or later).

Jazz Theory Quiz (http://jazzappsmobile.com/?page_id=44) Designed to test knowledge of music theory; broken down into five categories; can be of value to all musicians regardless of the type of music they play; quizzes cover note functions, chords, diatonic chords, scales, and intervals; (iOS 5.0 or later).

Jellybean Tunes (http://www.jellybeantunes.com/) Introduction to playing, reading, and creating music for all ages; includes different children's and classical songs; sing along, record, and play back performances; learn about the basics of written music through six narrated and animated lessons; create own songs with the intuitive editor; introduces children to written music; big, colorful notes, buttons, and backgrounds present music in a fun way suited to younger hands; playing songs by touching notes on the staff shows children where the notes are, what sounds they make, and what letters are associated with them; lessons present basic music theory; explore and create music; can take to a real piano or other instrument in place of sheet music; as the notes scroll by, user plays melody without having to stop to turn a page; replay feature allows user to listen to recorded performances; (iOS 4.3 or later).

Kids Music (http://www.123kidsfun.com/) Music game for toddlers and preschoolers ages one to four; includes twenty-four virtual instrument sounds including xylophone, drums, guitars, trumpets, flute, bells, and more; tap to hear the sounds; bright, colorful, child-friendly design; hundreds of rich graphics, vibrant sounds, and special effects; (iOS 5.0 or later).

playpad pro. Music Theory Stave Instrument (http://musicaltrixstar.com) Learn and play music; record songs and export; fifteen backing tracks to play along with; transpose into C, F, B♭, E♭, and A; key signatures up to seven sharps and seven flats; five voices include piano, strings, guitar, organ, and electric guitar; treble and bass clef; play chords and melodies; listen to an iTunes song and play along; learn the relationship between lines and spaces by physically playing the distances as chords or melodies in treble and bass clef; play tunes onto the colorful stave and hear what they should sound like; touch sensitive; (iOS 6.0 or later).

ThumbJam (http://thumbjam.com/) Musical performance experience; includes more than thirty high-quality real instruments multi-sampled exclusively for the app; hundreds of included scales allow user to play in any key and style; pick an instrument and jam; for soloing on an instrument; developed and tested by musicians to feel and sound like a real instrument; videos and audio at website; makes use of tilt and shake to add vibrato, tremolo, note bends, and volume swells for more realistic and expressive results; supports simultaneous touches and up to twenty-four voice polyphony; delay and stereo reverb add depth to the sound; resonant low-pass filter; build own loops from the ground up or import favorites; layer multiple loops with different instruments for each; export loops as audio files or DAW sessions via WiFi; copy and paste audio from other apps; supports both Audiobus and JACK to stream audio between other applications; broadcast tempo, key, and scale to other nearby devices via Bluetooth; can use as a MIDI or OSC controller over WiFi with any Core MIDI compatible device to control software instruments on a computer, control other apps on the same iOS device with Virtual MIDI, or play with other hardware or software MIDI controllers or apps; large variety of scales from Western to Eastern; can import new scales in the Scala format; record own instruments using a built-in or attached microphone; included instruments: cello, violin, viola, upright bass (plucked and bowed), several drumkits, darabukka, djembe, electric guitar, electric bass, trumpet, trombone, tenor sax, flute, low whistle, acoustic guitar, hammered dulcimer, mandolin, cittern, grand piano, drawbar organ (two types), rhodes, pipe organ, synth, strings, synth choir, theremin, sawtooth waveform, sine waveform, triangle waveform, and from Jordan Rudess, JR Zendrix, and JR Smooth Steel; many more for immediate download in the app including several drumkits, Mellotron, basses, banjo, tabla, ukulele, guitars, and more; new instruments posted regularly; (iOS 4.3 or later).

SONGWRITING APPS

American Songwriter (http://www.americansongwriter.com) Published bimonthly; in-depth interviews with up-and-coming, established, and legendary songwriters; articles on technology; reports on the business of music publishing with interviews and insights from publishers, producers, and industry professionals; explores all genres of music; (iOS 6.0 or later).

ASCAP (http://www.ascap.com/mobile/) Member Access; view and edit contact information; search and view entire ASCAP catalog of works; view and download royalty statements; view news headlines in an interactive photo slider; read a stream of important music industry news curated by ASCAP; ACE Database with detailed writer and publisher information for millions of musical works; contact publishers directly from the app; (iOS 5.0 or later; Android 2.1 and up).

Blues Writer (http://www.blueswriterapp.com) Writing, recording, and sharing app for songwriters, singers, and general musicians; enhanced notepad containing *Schirmer's Complete Rhyming Dictionary,* an industry standard including more than 96,000 rhymes; select from the ten professionally recorded blues tracks included in the app; record a song live; (iOS 3.1 or later).

BMI Mobile (http://www.bmi.com/faq/category/bmi_mobile) Can only be used by BMI-affiliated songwriters, composers, and copyright holders; mobile resource for career-building information and self-service options; manage BMI Live Performances, including set lists and bands; locate BMI Live Performance venues using GPS functionality; search Works Catalog for information on songs registered with BMI; paperless statement enrollment; set up direct deposit transactions; review royalty statements and tax documents with AirPrint capability; receive alerts of new royalty statements using push notifications; FAQs on all functions; payment date schedule; contact information for all BMI regional offices; image galleries and videos of current BMI events; career advice articles; (iOS 5.0 or later; Android 2.2 and up).

Figure (http://www.propellerheads.se/products/figure/) Make songs with drums, bass, and lead synth; play by sliding finger across the play pad; bass and lead parts use Reason's Thor synthesizer; drums powered by Reason's Kong drum machine; play in different keys and modes; set once or change on the fly; increase the Shuffle to loosen up beats; turn up Pump to add a club sound to tracks; adjust levels using Propellerhead's mixer; save, browse, and load song files; set length of loop; export audio to iTunes File Sharing; SoundCloud sharing; share tracks with access to Facebook, Twitter, and more; Audiobus supported input; supports AudioCopy and General Pasteboard; (iOS 6.0 or later).

Freestyle (http://www.musicroom.com/songwriting-apps) Writing, recording, and sharing mobile app for songwriters and poets; records vocal over the beat; sharing features; enhanced notepad containing *Schirmer's Complete Rhyming Dictionary*; record rhymes over a selection of exclusive beats included in the app; record over tracks; (iOS 5.0 or later).

How to Write Songs—Learn How to Write a Song Today! (Maurice Culbreath) Songwriting lessons, tips, and advice; song structures; writing hooks; songwriting ideas; (iOS 5.0 or later).

LyricFind (http://www.lyricfind.com/iphone/) Search by artist, song name, or lyrics; bookmark favorite lyrics; view artist biographies, photos, and album art; view charts of top lyrics and artists; (iOS 5.1 or later; Android 1.1 and up).

MetroLyrics (http://metrolyrics.com/apps.html) Lyric videos; millions of tracks; synchronize lyrics; find lyrics to favorite songs; use Map Charts feature to see the most popular songs locally and around the globe; join a worldwide music community; (iOS 4.3 or later).

MOGUL Songwriting & Recording Studio with Free Music Beats (http://www.makehitmusic.com/mogul.php) All in one studio app for iPhone and iPad that allows user to import beats, write lyrics, record vocals, and share music; (iOS 6.0 or later).

My Lyric Book (http://dctsystems.co.uk/) Have and carry all lyrics all the time; create set lists of songs that can be edited and reordered; share with the rest of the band; (iOS 5.1 or later).

Pocket Songwriter (http://www.musicroom.com/songwriting-apps) Writing, recording, and sharing app for songwriters, singers, and musicians; designed to aid and assist the songwriting process; enhanced notepad containing *Schirmer's Complete Rhyming Dictionary*; select from fourteen professionally recorded tracks and record a song; share online; (iOS 4.0 or later).

RhymeNowHD (http://www.purpleroom.com/ipad-rhyming-dictionary/) App is a rhyming dictionary for iPad; find rhymes with a database of 55,000 English words; (iOS 3.2 or later).

Rhymulator (http://www.paragoni.com/rhymulator-rhyme-app-for-iphone-ipad/) App is a rhyming dictionary and more for songwriters, rappers, and poets; (iOS 5.0 or later).

Rockmate (http://www.fingerlab.net/website/Fingerlab/Rockmate.html) Compose, play, record, and produce songs alone or with up to four musicians; complete music studio; smart chord progressions; separate guitar effects and audio samples produced in-house; customizable with multiple drums and keyboard styles; smart metronome; real-time looper; share songs by e-mail, Facebook, SoundCloud, or iTunes Folder Sharing; supports AudioCopy and General Pasteboard; Apple Core MIDI compliant; (iOS 6.0 or later).

Set List Organizer (http://usa.yamaha.com/products/apps/set_list_organizer/?mode=model#page=5&mode=paging) App can be used to edit and display the set list for a live performance; Apple Core MIDI compliant; (iOS 6.1 or later).

Song Board (http://www.foriero.com/en/pages/song-board-page.php) Create own orchestra at home or in school; includes six musical instruments and nine popular and well-known songs; connect iPads together via WiFi and play as a band; practice favorite songs and prepare a class band; create own melodies to play on the colored board; (iOS 6.0 or later).

Song Box—Lyrics Writer (http://www.songboxapp.com) App is designed for songwriters who need to have quick access to their song lyrics; (iOS 6.0 or later).

Song Writer (http://www.songwriterapp.com/www.songwriterapp.com/Welcome.html) Capture song ideas anywhere; write lyrics and chords; record a sample of the song; save in song list; continue work on computer or print out with e-mail option; (iOS 4.2 or later).

Songkick Concerts (http://www.songkick.com) Tracks tour dates and festivals for favorite artists and bands; (iOS 7.0 or later).

Songpen (http://loneyeti.com/songpen/) Journal for songwriters; Quick Lyrics and Quick Ideas features to jot down words and inspirations; keep track of lyrics, chords, notes, time signature, tempo, and key; record audio to keep track of melodies, instruments, or demos of the complete song; Inspiration feature gives a random tip, a random catchphrase, and a random picture; e-mail lyrics to self or collaborator; access recordings from device; (iOS 7.0 or later).

SongSynth (http://songsynth.wordpress.com) Portable songwriting studio; create and store song ideas; experiment with chords; record ideas; overdub vocals and melodies; (iOS 4.0 or later).

Songwriter's Pad (http://songwriterspad.com/songwriting-software-app-ipad/) App is used for generating song ideas; rhymes, dictionary, and thesaurus; chord notation; notes; author and publisher; audio recorder with backing tracks; sync songs; (iOS 7.0 or later).

Songwriter's Pad Multitrack (http://songwriterspad.com/songwriting-software-app-multitrack-recording-studio/) Pro songwriting tools; multi-track recording studio; beats; (iOS 6.0 or later).

SongWriting (http://www.moodworks.com/#!songwriting/c1404) Turn a germ of an idea into a finished song with lyrics, chords, and verse structure; listen and share; (iOS 5.1 or later).

StarComposer (http://p4spublishing.com/starcomposer/) Create songs with a few taps; original studio-quality music loops; record singer's voice on top; share animated music videos on YouTube and Facebook; create songs in two different musical styles, with eight more styles available for in-app purchase;

styles include rock, pop, dance, RnB, funk, metal, dubstep, drum and bass, blues, and reggae; available in multiple languages; (iOS 6.1 or later).

VCVCBC: The Ultimate Songwriting App (http://okeedoke.com/vcvcbc) Songwriter's tool; start from any element including lyrics, melody, chords, or rhythm; rhyming dictionary with Internet connection; take lyrics into song mode to add chords, melodies, and rhythms; Melody: slide finger up and down the keys to find the pitch then set a key and scale based on the last note played; highlights the notes used; Chords: start with more than twenty sets of common chord progressions and use any chord; play chord set like a guitar with built-in guitar strings and frets; play through chord progression or song live; Rhythm: tap out rhythms on common surfaces and record ideas with or without a metronome; Songs: combine lyrics and chords into a chord chart; add melody or rhythm ideas in the margin; rearrange, duplicate, and remove parts of the songs; switch to full-screen mode to play live; e-mail PDF of chord chart; undo/redo; type notes to self on the back of napkins; auto correct on/ off for lyrics; switch between all chords or use common chords; sort ideas by date or title; filter ideas by type; tutorial video included; songwriting aid; (iOS 5.0 or later).

VoiceJam (http://www.tc-helicon.com/products/voicejam/) App can create an entire song with user's voice, simulating bass lines, melodies, rhythms, and sound effects, all in one pass; audio looper; supports AudioCopy and General Pasteboard; (iOS 6.1 or later).

Recording Music: "The Sounds" Apps

BEATS, LOOPS, AND DRUM MACHINE APPS

Annoydio (http://www.legendofdopeyhollow.com/#) Grid of pads; assign, manipulate, and play sounds; create custom sounds using device's microphone; add custom sounds through iTunes File Sharing; manipulate the sounds by changing the frequency, playing in reverse, looping, or adding effects such as distortion and delay; change parameters such as attack, decay, sustain, and release; connect a MIDI keyboard and play the current sound with full polyphony; supports Inter-App Audio as a sound module and MIDI device; (iOS 6.1 or later).

Audio Elements (http://www.waveforms.pl/ae/) Gives user access to a large collection of professional audio loops including drum loops, basslines, synthesizer parts, special effects, and more that are ready to be used in music projects; browse the library and copy/paste selected sounds to another music app; content useful in any kind of looper, groove box, DAW, or sampler app like GarageBand, Akai iMPC, Loopy HD, Cubasis, BeatMaker2, Sunrizer Synth, Reactable, Electrify, and many more; supports AudioCopy and General Pasteboard; (iOS 6.0 or later).

Beat Shuffler (http://www.bigrobotstudios.com/#beatshuffler) Slices loops and allows user to trigger the slices; beat sketching tool where the slice triggering always falls on the beat; six tracks for loops; ability to import WAV and MP3 files; iOS File Sharing import for exchanging files with the app; mixer; master channel with effects; separate channel for each of the tracks with effects; available effects: delay and low-pass filter; AudioCopy support to import loops from other apps; Audiobus support to stream the audio output directly into other Audiobus-compatible apps; Audiobus supported input; (iOS 4.3 or later).

Beat Twirl (http://www.tivstudio.com/node/36) Beat slicer app; analyze existing sounds and detect rhythms with note onset algorithms; work with drum loops and rhythm patterns; enhance with additional percussion sounds or create new patterns; change the tempo of drum loops without changing the pitch; analyze percussion patterns and find the underlying timing; export single beats from a sound file; supports AudioCopy and General Pasteboard; (iOS 5.0 or later).

Beat Vibe (http://www.tivstudio.com/node/1) Programmable vibrating metronome/beat box with a flamenco clock display; practice tempo change and share rhythms; supports AudioCopy and General Pasteboard; Audiobus supported input; (iOS 6.0 or later).

Beat-Machine (Primitive Digital Software) Retro themed drum computer; combines X0X style sequencer with a sampler for control over beats; create and save patterns; supports AudioCopy and General Pasteboard; Apple Core MIDI compliant; MIDI Clock Sync; (iOS 5.1 or later).

BeatMaker 2 (http://intua.net/products/beatmaker2/) Composition interface for amateurs and professionals; virtual home studio; create multiple instruments and connect them to audio effects; compose, record, arrange, and transform ideas into complete songs to be shared with others; import sounds or use the sound library provided; more than 170 high-quality instrument and drum presets from synthesizers to orchestral sounds; import songs from iPod library; connect MIDI accessories for external control of instruments, effects, and mixer; for home-studio productions and live performances; compatible with audio interfaces; Universal app; MIDI Clock Sync; MIDI Support; supports iOS Inter-App Audio; MIDI Learn; Virtual MIDI port; Audiobus supported input, output; Apple Core MIDI compliant; (iOS 5.0 or later).

BeatMakerHD (http://www.createlex.com/apps) Drum machine with 128 trigger pads; sixteen pads over eight banks; control sound parameters on individual mute; record multiple audio tracks simultaneously; support for disk streaming to load samples without consuming RAM; share music with Sound-Cloud; record, draw, arrange, and pattern along the timeline to build beats; loop tempo editor; Sound Back included; native iPhone interfaces; Retina display; transfer files via iTunes and upload sounds; Audiobus supported input, output; (iOS 7.0 or later).

BeatStudio (http://frontierdesign.com/BeatStudio/AppStore) For playing, recording, and editing percussion sounds; use a custom drum set, build up patterns with multi-track recording, or change and adjust beats with the precision editor; intuitive interface for beginners and experienced players; performance mode lets users map a set of sounds from a built-in library of thirty percussion instruments to record a track; tracks can be entered free form, with a metronome, or along with a backing track imported with iTunes, WiFi, or Audio-Paste; multi-track mode to allow looping, editing, and layering of multiple percussion tracks; flexible tempo and quantization settings control alignment for recording and editing; full-screen editor provides the ability to fine-tune tracks; supports AudioCopy and General Pasteboard; (iOS 4.0 or later).

Beatwave (http://collect3.com.au) Comprehensive range of instrument samples and ways to mix them; Kaleidoscopic Beat Visualizer; samples of garage, hip-hop, funk keyboard, and synth sounds; mix the levels of tracks; add effects such as reverb; variable pitch control; adjustable tempo settings; Layers feature for up to four separate tracks allowing multiple instruments playing at once; tune sharing; enhanced visuals; SoundCloud support; (iOS 3.0 or later).

bip (http://soh.la/bip.html) Creative tool for playing, recording, and performing musical rhythmic patterns; touch and hold gestures are used to select beats and patterns; patterns can be mixed and matched along with a touch-strip controller for transforming beats; choose from track templates for creating tracks or import and tweak beats with own samples; hours of creative composing and live performance; simple and clean interface; all beats can be programmed to play in time including complex subdivisions; touch strip for changing parameters of any beat, loop, or entire track; record beats and parameters to create expressive musical patterns; arrange patterns on the fly for each track; use to compose, arrange, and perform live; load track presets for quick music making; more than two hundred samples included; import samples using AudioPaste or iTunes File Sharing; Audiobus supported input slot; (iOS 7.1 or later).

DM1 The Drum Machine (http://fingerlab.net/website/Fingerlab/DM1 .html) Advanced vintage drum machine; turns iPad into a fun and creative beat-making machine; easy and fast to use; loaded with ninety-nine electronic drum kits and hyper-realistic graphics; mix between fun playability and powerful sonic capacities; twenty-eight classic vintage drum kits; seventy-one in-house produced electronic kits, edited and mastered at Fingerlab professional studio; offers five main sections: (1) the Step Sequencer uses the multi-touch screen to turn on and off steps in sequence with fingertips to create beats or unexpected rhythms; (2) the Drum Pads let user play and record the beat with automatic quantizing and pitch-bend ribbon; (3) the Mixer for sound mixing of drum kits featuring settings for volume, pitch, sample length, custom drum kit element for each channel, mute, and solo mode; (4) the FX Trackpads to distort, modulate, and transform beats with effects including Overdrive, Reverb, Delay, Phaser, Formant, Texturizer, Robotizer, Filter, Dalek Filter, and Compressor; (5) the Song Composer allows user to make a song with the beat patterns created by dragging and dropping the patterns onto the timeline; Retina display; time signature; sixteen or thirty-two steps per pattern; samples import from iPod Library, microphone, Dropbox, AudioCopy, and iTunes shared folder; export to SoundCloud, Dropbox, Facebook, e-mail, iTunes shared folder, or AudioCopy; separated tracks and AAC encoding export options; the Randomizer tool; fast drum kit loading; playable pattern selection; Audio background mode; WIST synchronization technology; Audiobus support; full MIDI implementation; Inter-App Audio support; video trailer on YouTube; (iOS 6.0 or later).

Donut (http://thestrangeagency.com/ipad/#app-donut) Record loops in two dimensions; use like a phrase looper by recording horizontally while using step controls through vertical recording position; use as a real-time processor by using ThruRecord mode; record and play back at different speeds; record backward; record and play back diagonally across the surface, creating staccato sounds; record and play back with multiple heads, recording or playing multiple positions at once; supports AudioCopy and General Pasteboard; (iOS 5.1 or later).

DopplerPad (http://www.dopplerpad.com) Create and perform musical hooks, phrases, and loops with a variety of custom synth and sample-based instruments; make samples on the fly and weave them into compositions; expressive touch instrument; use rhythmic and melodic automation to create phrases by moving fingers around the musical touchpad; thirty-seven instruments including leads, bass, sound effects, drums, and samplers; two musical touchpads; visual feedback to see the groove; two recording modes: overdub and replace; mixer for cross-fading, muting, and playback control; save loops and phrases; load them into the performance; gate arpeggiator to create expressive rhythmic patterns; create samples anywhere with external or built-in mics; adjustable tempo and loop length; programmable scale to stay in key; adjustable octave for a full range of notes; supports AudioCopy and General Pasteboard; (iOS 3.1 or later).

Droboto (http://makeaudioapps.com) Live music performance app that provides sixty-four robotic distorted drum samples on drum pads; (iOS 7.0 or later).

Drum and Beat (http://www.mobobo.com) Make beats, scratches, samples, and music; includes a fully featured beat box, drum machine, sampler, scratch pad, and music sequencer; beat box: select or unselect the beats for each track; use any sample or own recordings; save beats, load them back, or share them; scratch pad: scratch like a DJ; use any sample or own recordings; automatic pitch adjust and manual or auto drive system; save scratch sessions, load them back, or share them; drum machine: play the drums; use any sample or own recordings; save drum sessions, load them back, or share them; sampler: record samples with the microphone; adjust pitch, speed, and reverb; save samples, load them back, or share them; sequencer: use any sample or own recordings; add, edit, and remove tracks; adjust timings of samples and listen to music; save music, load it back, or share it; create, play, edit, save, and load sessions for free; use own recordings via the sampler; use beats, drum sessions, samples, and scratches as instruments; share with friends; full-screen Retina support; real-time 3D graphics; optional in-app purchase; extra tracks in the beat box; record longer scratches, drum samples, and samples; additional sequencing tracks; more than one hundred additional sound samples; (iOS 4.3 or later).

Drum Beats+ (http://www.ninebuzz.com/beatsplus/) Collection of more than one hundred drum loops in various styles; expandable to more than 250; for jamming, songwriting, practice, and performance; loops range from one to four bars in length; for casual practice or exploring song ideas; plug into speakers for the best sound or play from device; includes Hit Song Pack of remakes of beats from sixteen chart-topping songs; easy tempo changer from 60 to 190 bpm with intervals of 5 bpm; tempo count-off; professionally mastered sound; favorites; randomizer button; use as a metronome for rhythm practice; (iOS 6.0 or later; Android 2.2 and up).

Drum Machine (http://trajkovski.net/sound.html) Studio-quality sound; eleven popular drum kits; eight drum pads; six sound effects; multi-touch; pad pressure; pad aftertouch; integrated sampler; integrated sequencer; integrated mixer; integrated recorder; sensor pitch bend; sensor volume change; drum machine editor; samples recording; samples load and save; tracks load and save; machines load and save; MIDI load and save; MIDI over WiFi; MIDI velocity; MIDI aftertouch; MIDI volume control; MIDI pitch bend; twelve sounds polyphony; seventeen machine skins; twenty-one pad styles; eighteen text styles; seven pad fonts; eight window themes; window animations; custom pad text; multi-window support; (Android 2.1 and up).

DrumJam (http://www.drumjamapp.com) Percussion and drums for the iPad, iPhone, and iPod touch; designed by award-winning international percussionist Pete Lockett and Sonosaurus LLC, the developer of ThumbJam; intuitive quantized beats performed by dragging fingers; build layers from a wide range of authentic ethnic percussion and drum kit loops; jam over the top of original grooves; includes dozens of different looped instruments to choose from, each with up to twenty different patterns; change the levels, alter the pan, mute, solo, and tweak a low-pass filter giving flexibility over each individual part; randomize all or some of the parts for unexpected changes; MIDI Clock Sync; Virtual MIDI port; supports AudioCopy and General Pasteboard; Audiobus supported input; Apple Core MIDI compliant; (iOS 5.0 or later).

DrumStudio (http://www.rollerchimp.com/drumstudio/) Mobile drumming app that lets user create realistic drum programming; includes advanced features for seasoned professionals like complex time signatures, flams, chokes, and Audiobus integration; easy to use for drumming novices; record, edit, play back, and export beats; loop any section or bars by using the slide out loop selector; get mobile beats into Logic, Pro Tools, or Cubase with the MIDI exporter; copy and paste tab files from websites; play back tracks at different tempos and time signatures; section editing for verse/chorus creation; mix 6/8, 5/4, 7/8, or any time signature in any song; multiple tempos in any song; play back to play along with any iTunes song; multiple bar copy and paste;

Audio copy and paste; designed for both beginners and professional drummers; for drum students and teachers; MIDI Clock Sync; (iOS 5.0 or later).

DrumtrackHD (http://www.simple-is-beautiful.org) Drum step sequencer for the iPad; features include FX audio effects, up to sixteen tracks of scalable patterns with up to thirty-two steps, track automation, pads, signature modification, sample patterns, new HD drum kits, reverse samples, and more; create rhythms in intuitive interface using a choice of more than fifteen drum kits; customize kits by importing user samples, adding tracks, subtracting tracks, replacing tracks, or creating multi-sample tracks; add audio FX, swing, and more while playing back; export patterns or songs in General MIDI or WAV Format, transfer files with iTunes, and/or upload to SoundCloud; supports AudioCopy and General Pasteboard; (iOS 3.2 or later).

Dubstep Kit (http://www.dubstepkit.com) Create dubstep music, beats, and more; for enthusiasts and DJs; includes 384 free downloadable HD 192-kHz stereo audio files for commercial use; 100 percent royalty free; clear audio with Retina graphics and a full-screen mode; (iOS 6.0 or later).

Earhoof (http://www.psicada.com) Virtual instrument; turns iPad into a rhythmic gradient of sound; sophisticated instrument creation tool; layer multiple audio files and combine them with high-quality digital signal processing; innovative internal rhythm engine; switch seamlessly between complex rhythms while remaining in control of the generated sounds; (iOS 7.0 or later).

EasyBeats (http://easybeats.net/index.html) Create new four-bar beat patterns; use the in-built samples or imported sounds; includes sixteen responsive sample pads; sample kits; four loopable bars each sixteen step quantized (sixty-four step total); multi-touch pattern editor with drag/drop and copy/paste support; ability to change individual samples to create new kits; alter individual sample trigger velocity for each step of a pattern; import own samples; import kits from Intua's Beatmaker via local BeatPack server; (iOS 6.1 or later).

Electrify (http://www.electrify.info) Virtual sample-based Groovebox for iPad; Audiobus supported input; Apple Core MIDI compliant; (iOS 5.0 or later).

Everyday Looper (http://www.mancingdolecules.com/everyday-looper/) Record musical phrases and loop them immediately; mix and merge; metronome; auto start/stop quantized recording; tools and options menu; supports AudioCopy and General Pasteboard; (iOS 5.1 or later).

FingerBeat (http://www.fingerbeat.com) Pocket sampler and drum machine; grab sounds using microphone; supports AudioCopy and General Pasteboard; (iOS 3.0 or later).

FunkBox Drum Machine (http://syntheticbits.com/funkbox.html) Designed to look, feel, and sound like a classic beat box from the 1970s and 1980s; uses original audio samples taken directly from a collection of vintage drum machines; hundreds of classic sounds; thirty-six funky preset drum patterns; tap along on top of the beat using the drum pads and overdub parts; use the built-in mixer to adjust volume, pan, and accent levels for each sound; save, load, and share beats; adjust setting of tempo with the tempo slider or tap a rhythm; adjustable swing; sync up to external drum machine or groovebox hardware with MIDI sync; MIDI Clock Sync; Virtual MIDI port; Audiobus supported input; Apple Core MIDI compliant iOS app; allows user to connect MIDI class compliant hardware to iOS devices via interfaces; (iOS 5.1.1 or later).

GlitchBreaks (http://www.glitchbreaks.com) Universal manipulation tool for glitching breakbeats; manipulate beats to create new beats and record them; Audiobus supported input, effects; supports AudioCopy and General Pasteboard; (iOS 5.1 or later).

GrooveMaker 2 (http://www.ikmultimedia.com/products/cat-view.php?C= family-groovemaker) App for creating non-stop electronic, dance, and hip-hop tracks in real time, by anyone, anywhere, like a professional DJ; more than 120 loops included; (iOS 6.0 or later).

iMaschine (http://www.native-instruments.com/en/products/maschine/ maschine-for-ios/imaschine/) User can create a beat on sixteen pads; add a melody with the keyboard; use a field recording or lay down vocals; sample directly from iTunes; (iOS 6.0 or later).

iMPC (http://www.akaipro.com/product/impc) More than 1,200 samples, fifty editable programs, and eighty editable sequences; exclusive iMPC sound set and classic Akai Professional sound library; sample from iPad mic, line-in, or music library using multi-touch turntable interface; export tracks to MPC Software for use with MPC Renaissance and MPC Studio; share on SoundCloud, Facebook, Twitter, Tumblr, or export to computer; Sixteen Levels mode: sixteen levels of attack, velocity, length, and tuning; Note Variation slider: adjust attack, velocity, length, and tuning while playing; trim and categorize samples, undo takes, and erase samples; create new sequences; record and overdub beats; Live Sequence queuing for real-time performance and playback; drag and drop program creation; built-in effects: delay, bit crusher, master compressor/limiter; Time Correct with variable swing; Note Repeat; supports WIST; AudioCopy beats and sequences to compatible apps; AudioPaste samples from compatible apps; Tabletop Ready iMPC is free for iMPC owners; route to and use with Tabletop Effects and devices; Line-in allows user to resample Tabletop devices; use with Tabletop's Timeline Editor; supports iOS Inter-App Audio; (iOS 6.1 or later).

Keyboard Arp & Drum Pad (http://usa.yamaha.com/products/apps/keyboard _arp_and_drum_pad/?cnt=iT) Core MIDI application that transmits notes to external MIDI instruments; includes a keyboard, arpeggiator, and assigable drum pads; Apple Core MIDI compliant; Virtual MIDI port; (iOS 5.1 or later).

KORG iELECTRIBE (http://www.korg.com/us/products/software/ielec tribe_for_ipad/) Korg's first dedicated app; analog-synth style beat making; takes full advantage of iPad's multi-touch display; new style of musical instrument; virtual analog beat box; choose a part or sound and touch the sixteen-step sequencer to build a groove; re-creates the historic sound engine and sequencer capabilities; provides advanced motion sequencing, eight effect types, and sixty-four new preset patterns; create music in a wide variety of dance and electronic music styles; Sound Engine: Analog Synthesizer Modeling; PCM Samples; Virtual Valve Force Tube Modeling; Instrument Voices/ Parts: eight total including four percussion synthesizer parts and four PCM synthesizer parts; Effect Section: master effect assignable per step/per part; eight effect types: Short Delay, BPM Sync Delay, Grain Shifter, Reverb, Chorus/Flanger, Filter, Talking Modulator, and Decimator; Sequencer Section: patterns include sixty-four steps maximum per part; motion sequencing can memorize all knob motions in a pattern; Tempo: 20 to 300 bpm with Tap tempo and Swing function; Pattern-set function; no Song function; Pattern Memory: 160 total; audio export function including bouncing a pattern and real-time recording of performances in 16-bit/44.1-kHz stereo WAV format; publish and share an exported audio file online with the SoundCloud audio platform; Virtual MIDI port; (iOS 5.1 or later).

KORG iELECTRIBE Gorillaz Edition (http://www.korg.com/us/products /software/ielectribe_gorillaz/) Korg teamed up with the globally successful virtual band Gorillaz to create a new dedicated beat box for the iPad; Audiobus supported input; Apple Core MIDI compliant; Virtual MIDI port; (iOS 5.1 or later).

KORG iKaossilator (http://www.korg.com/us/products/software/ikaossila tor/) Play sounds by stroking, tapping, or rubbing the screen; visual effects move along with performance; 150 built-in sounds cover any style of dance music; scale/key settings; create tracks with loop sequencer using up to five parts; control loops in real time for DJ-style live performances; Audio Export records and saves performance; SoundCloud allows users to share and remix loops with other users; WIST capability allows wireless sync-start with other music apps; AudioCopy/AudioPaste Support; Audio Loop Import; Audiobus supported input; Virtual MIDI port; seamless operation in tandem with the Kaossilator 2; (iOS 5.1 or later).

Loop Twister (http://www.waveforms.pl/code_recent.html) Intuitive beat remixing app; designed for playing and real-time processing of musical loops; loaded with ninety-six loops and DSP effects; Audiobus supported output; (iOS 6.0 or later).

LoopMash HD (http://www.steinberg.net/en/products/mobile_apps/loopmash_hd.html) Innovative approach to mixing music; simultaneously merges up to eight four-bar loops by matching and replacing comparable elements; nineteen live performance effects such as tape stop and stutter effects; over thirty presets and 258 loops from ethno to electro; stripped-down version of the LoopMash virtual instrument featured in Cubase; four studio-grade effects for more sound control; Audiobus supported input; intuitive 3-D swipe-page navigation; (iOS 6.0 or later).

Loopr Live Loop Composer (http://loopr.net) Record and play loops; create, practice, jam along, or sample audio soundscapes; ergonomic and simple; one-finger controls or hands-free MIDI pedalboard compatibility; Apple Core MIDI compliant; MIDI Clock Sync; (iOS 7.0 or later).

LoopTastic HD (http://www.soundtrends.com/apps/looptastic_hd) Create remixes and compositions; drag and drop loops, mix DJ-style, and add effects; supports AudioCopy and General Pasteboard; Apple Core MIDI compliant; MIDI Clock Sync; (iOS 3.2 or later).

LoopTastic Producer (http://www.soundtrends.com/apps/looptastic_producer/) Create non-stop mixes and new music tracks; live remix and loop-based performance tool; supports AudioCopy and General Pasteboard; Apple Core MIDI compliant; MIDI Clock Sync; (iOS 3.0 or later).

Looptical (http://moocowmusic.com/Looptical/) Complete music production solution; record musical ideas on the go and mix them into a finished track or use them as the basis for a new project back at the studio; create looped sections of music and combine them to form larger musical sections to create a finished song; comprehensive effects and fully automated mixer controls; export MIDI and component audio tracks to DAW when back at the studio to take song one step further; supports iOS Inter-App Audio; (iOS 6.0 or later).

Loopy HD (http://loopyapp.com) Layer looped recordings of singing, beat-boxing, or playing an instrument; interactive tutorial system; Apple Core Audio; supports AudioCopy and General Pasteboard; Audiobus supported input, output; MIDI Clock Sync; MIDI Learn; Apple Core MIDI compliant; Universal app; (iOS 5.0 or later).

LP-5 (http://www.finger-pro.com/lp-5.html) Performance-geared loop player for live arranging and remixing; trigger audio loops using an intuitive launch matrix while automatic tempo synchronization keeps in time with the beat; Apple Core MIDI compliant; MIDI Clock Sync; (iOS 6.1 or later).

Mad Pad (http://www.smule.com/apps#madpad) Turn everyday sights and sounds into the ultimate percussive instrument; (iOS 4.2 or later).

MIDI Loop Station (http://www.MIDIloopstation.com) App for the performing musician; specify loop boundaries in a MIDI file and supplement each loop with sixteen tracks; can be remote controlled by a MIDI controller via Core MIDI; starting from scratch, gradually build up the sound by playing and adding instruments; (iOS 4.3 or later).

mobileRhythm mR-7070 (http://www.mobilerhythm.co.uk) Classic drum machine for iPhone and iPad; 1980s look and sound; MIDI in, out, and MIDI clock out; precise tempo control with digital readout; banks for pattern storage; individual faders for each drum sound; multiple beat divisions for time signatures with adjustable last step; solid timing; grid-view pattern display; hardware drum machines and beat boxes from the 1980s have defined musical genres from house, acid, hip-hop, to the latest dubstep, grime, and other electronic beats; works like the original beat boxes; easy to program complicated beats; MIDI settings are in the iOS settings app; Apple Core MIDI compliant iOS app; allows user to connect MIDI class compliant hardware to iOS devices via interfaces; (iOS 4.0 or later).

MoDrum Rhythm Composer (http://www.finger-pro.com/modrum.html) Beat-making app that features real-time synthesis of drum sounds that range from TR-808 style to real; sound characteristics can be tweaked from authentic to affected; virtual analog drum machine with a thirty-two-step grid sequencer, sound-sculpting parameters, a tempo-matching delay effect, dynamic compression, reverb processor, sophisticated MIDI support, audio looper, performance recorder, and more; Audiobus supported input; Apple Core MIDI compliant; Virtual MIDI port; (iOS 4.2 or later).

Molten Drum Machine (http://www.onereddog.com.au/molten.html) Tool for creating percussive sounds and rhythm sequences for the iPad; based pattern sequencing that divides time to add tuplets and fills to drum programming; built on a versatile synthesizer that allows user to play and process own sound samples; kits can be mangled, filtered, and crushed using the multitouch interface; load and store patterns and share them with friends; export to a standard MIDI file and import into a professional digital audio workstation; supports AudioCopy and General Pasteboard; Audiobus supported input; Apple Core MIDI compliant; MIDI Clock Sync; Virtual MIDI port; (iOS 5.0 or later).

My BeatBox (http://trajkovski.net/sound.html) Create automatic rhythms from samples of voice; includes twelve popular rhythms; one custom rhythm; integrated sixteen-step sequencer; variable speed (bpm); electronic sample set; acoustic sample set; human sample set; user sample set; samples recording; samples importing (long press); sample set save and load; sequences saving; samples mix levels; samples tuning; global tuning; four-channel effect; fifteen backgrounds; tablet/pad support; hardware acceleration; (Android 2.1 and up).

Rhythm Pad (http://www.rhythmpad.com) Play drums on mobile devices; includes a wide range of high-quality real and electronic drum kits; connect device to speakers and start jamming; can also play along with songs from iTunes library; created by musicians for musicians; realistic open and closed hi-hats function; pads sound louder at the center; optimized layout for maximum playability in iOS devices; stereo sampled sounds with maximum polyphony give a realistic feel; ability to configure the levels for individual pads; added ability to rearrange pad positions; record patterns and share recordings; Audio-Copy: allows user to take samples recorded in app to other audio applications; MIDI support: sends MIDI via WiFi or Camera Connection Kit allowing user to use app as a MIDI controller; while sending MIDI signals, drums are mapped based on General MIDI standard; Audiobus support: stream audio from app to other Audiobus-compatible apps; kits include Real, Real with Ride, Real with Rimshot, Real Pop Kit, Dance/Club, Chemical, Hip-Hop, Percussion, Rock Kit, Trip-Hop; Stickteck's Drummerz tablet drumsticks make playing app more realistic; for practicing, jamming, and recording favorite drum tracks from anywhere; (iOS 7.0 or later).

Robotic Drums (http://roboticdrums.com) Analog modeled drum synth with a probability sequencer; combination makes the app suitable for live sound manipulation, sequence experimentation, and regular beats; minimalist interface with a high emphasis in flow; user controls how random a sequence is; quantized pattern switching; Audiobus support; MIDI out messages when voices trigger; MIDI in changes patterns via note MIDI messages; MIDI Clock Sync support; eight analog modeled synthesized voices, each with a noise and an oscillator section; flexible configuration allows each voice to generate sounds including punchy bass drums, snappy snares, lo-fi tones, and sound effects; sequencer for each voice is based on probabilities; each session can hold sixteen patterns; switching patterns in a live situation is controlled by a pattern quantization parameter allowing for pattern switches to occur exactly when expected, maintaining the musical flow of a piece; Dropbox support for saving sessions and sharing across devices; (iOS 6.0 or later).

S4 Industrial Composer (http://www.sobal.co.jp/s4/archives/1326) Beat box allows user to compose and play sixty-four industrial beat patterns; can also record the patterns; Audiobus supported input; (iOS 5.0 or later).

S4 Rhythm Composer Pro (http://www.sobal.co.jp/s4/archives/2097) Beat box that can be customized in various ways; fourteen drum kits are preset; real-time effects; create and save; allows user to compose sixty-four beat patterns; supports AudioCopy and General Pasteboard; Apple Core MIDI compliant; Audiobus supported input; (iOS 5.0 or later).

Sector (http://kymatica.com/Software/Sector) Slice and sequence loops with a matrix of markov-chain connections; introduce order with the probability based coin-flipping pattern sequencer; bend time with warp functions for glitches and modulations; tweak and trigger in real-time performance; from rigid control to random chaos; (iOS 5.1.1 or later).

SessionBand (http://sessionbandapp.com) Chord-based audio loop app; more than eighty thousand copyright-free audio loops; supports AudioCopy and General Pasteboard; (iOS 5.1 or later).

Stochastik Drum Machine (http://stochastik.xitive.com) Full-featured drum machine; can set a probability that a note will trigger for each step; basic beat can rock steady while accents and syncopation happen occasionally; generate thousands of unique variations automatically in the time it takes to sequence one, two, or four bars; as easy as programming a static loop but generates a dynamic track without repetitive beats; over a thousand high-quality samples are included, from classic analog drum machines of the past to acoustic jazz and rock kits; mix and match samples to create the perfect kit for each song; import own samples through iTunes File Sharing, e-mail, or Dropbox; plays well with other apps; bounce pattern to a WAV file, and it's automatically put in the clipboard for pasting into other apps; wirelessly sync with other devices; with background audio can lay down the beat while using a synth; supports AudioCopy and General Pasteboard; (iOS 7.0 or later).

StompBox (http://www.4pocketsaudio.com/product.php?p=6) Turns iPad into a digital effects unit complete with a chromatic guitar tuner, a metronome, and loop recording tools; Apple Core Audio framework; Audiobus supported input, effects; (iOS 5.1 or later).

TNR-i (http://usa.yamaha.com/products/musical-instruments/entertainment/tenori-on/tnr-i/?cnt=iT) App is the iOS version of the world-renowned Tenori-on; supports iOS Inter-App Audio; supports AudioCopy and General Pasteboard; (iOS 6.1 or later).

WerkBench (http://www.bolasol.com/werkbench/) Beat box that is part loop pedal, part drum machine; two sequencers that let user instantly sample sounds into any place in the rhythm and then alter them in real time; Audiobus supported input, output; (iOS 4.3 or later).

EFFECTS APPS

AD 480 Pro (http://fiedler-audio.com) Pro-quality studio reverb effect; use with Audiobus, Inter-App Audio, iTunes music library, and multichannel audio interfaces; record, process, and export directly; realistic acoustic sound; special graphics mode/stage mode for low-light environments; MIDI functions let user fully remote control and automate; audio routing matrix for flexibility in a multi-app/multi-channel environment; professional reverb sound effects with variable room size; 108 presets; professional preset management for MIDI, audio routing, and reverb settings; extensive filter layout for a wide spectrum of sound; supports internal and external USB class compliant or MFi audio interfaces; routing matrix supports up to twenty-four input and twenty-four output channels; full Audiobus integration (input, filter, and output position); supports full MIDI communication via network and USB-MIDI interfaces as well as inter-app MIDI; supports sample rates of 44.1 kHz to 96 kHz; latencies down to sixty-four samples; background audio for multitasking

with other apps; audio recording, playback, offline processing; audio import and export via AudioCopy/AudioPaste/AudioShare; (iOS 6.0 or later).

AmpKit+ (http://agilepartners.com/apps/ampkit/) Guitar amps, effects, and recording; Retina display; Gear Store with the first full-featured, modern bass amp for iOS; supports AudioCopy and General Pasteboard; Apple Core Audio framework allows user to use class-compliant audio interfaces; MIDI and AirTurn Controller support; Universal app; (iOS 5.0 or later).

AmpliTube (http://www.ikmultimedia.com/products/amplitubeiphone/) Real-time guitar and bass mobile multiFX and recording app; includes eleven stomps, five amps, five cabinets, and two mics; add more gear models via in-app purchase; single-track recorder expands to four-track studio with master effects via in-app purchase; AutoFreeze feature allows for separate effects rigs on each recorder track; Studio feature turns multi-track into a DAW with full editing features; loop drummer module that plays in sync with the recorder; Inter-App Audio support; Audiobus support; import and play songs as backing tracks from music library or computer; export via e-mail, SoundCloud, FTP, or File Sharing; Audio Copy/Paste; slow down/speed up tempo of imported songs without changing pitch; No Voice removes lead vocal or guitar solo from imported songs; management of presets and favorites; tuner/metronome; digital audio in/out; MIDI controllable; low latency for real-time playing; (iOS 6.0 or later).

AmpliTube Orange (http://www.ikmultimedia.com/products/orangeipad/) Turn iPad into a mobile Orange guitar/bass multi-FX processor and recording studio; use the iRig HD, iRig, or iRig STOMP interface to connect guitar to iPad; use iRig MIDI to control AmpliTube with floor controller; developed at Orange to ensure the authenticity of the sound; (iOS 6.0 or later).

AudioReverb (http://www.virsyn.net/mobileapp/index.php?option=com_content&view=article&id=11&Itemid=8) Studio-quality reverb for Audiobus/Inter-App Audio and iTunes music library; algorithmic reverberation app combining the flexibility of vintage algorithmic reverbs with the sonic quality of convolution-based reverbs; uses impulse responses from real acoustic spaces and classical reverb algorithms to emulate acoustical properties; professional reverb plug-in to improve the quality of any Audiobus or Inter-App Audio compatible app; smooth reverb tail generation; early reflection modeling; 118 factory presets; four band equalizer; Inter-App Audio compatible effect; Audiobus compatible as input, effect, output; reverberate any song in iTunes library; low-latency live usage with mic and headphones; audio recorder; upload to SoundCloud and File Export; Audio pasteboard for exporting to other apps; (iOS 5.0 or later).

AUFX:Dub (http://kymatica.com/Software/AUFX) Stereo delay audio effect; for clean digital delays to filthy dub echoes and broken cassette tapes; Audiobus supported input, effects, output; Apple Core MIDI compliant; MIDI Clock Sync; supports iOS Inter-App Audio; MIDI Learn; Virtual MIDI port; JACK Audio/MIDI Connection Kit supported; (iOS 5.5.1 or later).

AUFX:PeakQ (http://kymatica.com/Software/AUFX) Four-band parametric stereo equalizer; frequency enhancements; boosting/cutting; special resonator effects; (iOS 5.1.1 or later).

AUFX:Space (http://kymatica.com/Software/AUFX) High-quality real-time reverb audio effect, for small rooms, resonators, and infinitely large spaces; app is the first in the AUFX series, a collection of simple and powerful audio effect apps; Audiobus supported input, effects, output; supports iOS Inter-App Audio; MIDI Learn; (iOS 5.5.1 or later).

Bias (http://www.positivegrid.com) Amp designer, modeler, and processor; (iOS 7.0 or later).

Compressor (http://www.audioforge.ca/more.php) Multiband compressor brings audio track to life adding impact and character; upload music files via WiFi; supports AudioCopy and General Pasteboard; (iOS 3.0 or later).

Crystalline Shimmer Effects Processor (https://holdernessmedia.zendesk .com/hc/en-us/requests/new) Shimmer reverb/delay effect for Audiobus and Inter-App Audio; create synth-like tones and textures from almost any sound source; for guitar, vocals, synths, sound design, and more; (iOS 7.0 or later).

Echo Pad (https://holdernessmedia.zendesk.com/hc/en-us) Multi-effects processor; range of sounds including authentic dub echo, crystal echoes, multi-tap delay, chaotic self-oscillating analog delays, and bit crushed mayhem; Audiobus supported input, effects, output; supports iOS Inter-App Audio; (iOS 5.1 or later).

Effectrix (http://www.sugar-bytes.de/content/products/EffectrixIOS/index .php?lang=en) App is a professional multi-effect sequencer for the iPad; (iOS 7.0 or later).

Equalizer (http://audioforge.ca/equalizer.php) Equalizer; play iPod library; total playlist control; supports AudioCopy and General Pasteboard; (iOS 7.0 or later).

Fiddlicator (http://fiddlicator.wordpress.com) Audio tool that simulates various kinds of acoustic environments by the convolution of the input signal with a custom impulse response; developed to simulate an acoustic body resonance for electric musical instruments, mainly the electric violin; can be used for a cab simulation with proper impulse response files; (iOS 5.1.1 or later).

Filtatron (http://www.moogmusic.com/products/apps/filtatron) Real-time audio filter and effects engine; use sound from sound sources including line/microphone input, sampler, built-in oscillator; import any song on device into the sampler; supports AudioCopy and General Pasteboard; (iOS 5.0 or later).

Final Touch (http://www.positivegrid.com/finaltouch/) Audio post-prodution system; maximizer, equalizer, four-band compressor, stereo imager, reverb, and dithering; gives mixes a balanced, polished, and professional sound; (iOS 7.0 or later).

GrainProc (http://grainproc.e7mac.com) Provides an expressive control surface for granular manipulation of real-time audio input for sonic sculpting and self-accompaniment of guitar, voice, and more; quick control with fingers; Audiobus supported effects; (iOS 5.0 or later).

iDensity (http://www.densitygs.com) Designed for asynchronous sound file granulation; genuine granular playground able to generate a wide range of effects including time/pitch shifting, time/pitch jittering, intricate textures, grain fountain/pulverizer, recording and manipulation of buffers, complex scrub pad exploration, dynamic envelope shape, and more; Audiobus supported input, effects, output; supports AudioCopy and General Pasteboard; (iOS 6.1 or later).

iPulsaret (http://www.densitygs.com) Real-time software capable of all time-domain varieties of granular synthesis; genuine granular playground able to generate a wide range of effects including time/pitch shifting, time/pitch jittering, intricate textures, grain fountain/pulverizer, recording and manipulation of buffers, dynamic envelope shape, and more; basic and advanced tools needed to manipulate sound; optimized, user-friendly, visually sharp interface for fast improvising live, composing in a studio, or sound designing at home; load from built-in sound sample; record audio directly with iPad mic; add WAV, AIFF files via iTunes; capture the stereo output of granulation into the documents folder while playing the keyboard or moving granulation parameters in real time; Audiobus supported input, output, effects; (iOS 6.1 or later).

iShred LIVE (http://frontierdesign.com/iShredLIVE/AppStore) Effects and tools for guitar; amp simulator with variable overdrive; starts with two free effects, the HK-2000 digital delay and Q-36 Space Modulator/Flanger; built-in noise gate and filter; supports AudioCopy and General Pasteboard; (iOS 4.3 or later).

JamUp Pro XT—Multi Effects Processor (http://www.positivegrid.com/jamup/) Digital multi-effects processor; turns device into a studio-grade guitar rig; comes with one amp and six effects; supports AudioCopy and General Pasteboard; Apple Core Audio; Universal app; (iOS or later).

Level.24 (http://www.elephantcandy.com/app/level-24/) Pro-quality real-time spectrum analyzer, EQ, compressor, and limiter; up to 1/24 octave band analysis precision and twelve discrete parametric equalizers and ultra low latency; audio control on the iPad; fully Audiobus compatible to receive, process, and record sound from other Audiobus apps; (iOS 6.0 or later).

LiveFX (http://www.elephantcandy.com/app/livefx/) Dynamic effects processor for iPad; any audio source can be processed; use either live sound input, the built-in track player that plays from iTunes library, or Audiobus to stream audio through the app; create own combination of audio effects; simultaneously control up to four effects using a single fingertip; more than one thousand possible combinations; use multi-touch for automated dynamic sound effects; Audiobus supported input, effects, output; (iOS 6.0 or later).

Master FX (http://audio-mastering-studio.blogspot.com) Multipurpose effect processor for guitar, synthesizer, and voice; provides a wide range of high-quality and creative effects options for audio processing; (iOS 6.0 or later).

Reverb (http://www.audioforge.ca/feedback/index.php?board=1.0) Simulate small intimate rooms or large concert halls; add reverb or echo to recordings, voice samples, instruments, or complex samples or mixdowns; supports AudioCopy and General Pasteboard; (iOS 3.0 or later).

RoboVox (http://www.mikrosonic.com/robovox) Voice changer with twenty-four different voice effects; Audiobus supported input, effects; (iOS 7.0 or later).

Squashit (http://klevgrand.se/) Audio effect filter provides a unique distortion algorithm; doesn't sound vintage; developed by musicians; Inter-App Audio compatible; (iOS 7.0 or later).

Stereo Designer—Stereo Shaper and Mid/Side Processor (https://holderness media.zendesk.com/hc/en-us/requests/new) Multi-purpose stereo shaping tool for Audiobus and Inter-App Audio; "stereoize" a mono sound source, apply mid/side processing, independent high- and low-pass filters per channel, and more; (iOS 7.0 or later).

StompBox (http://www.4pocketsaudio.com/product.php?p=6) Multi-FX processor with three effects: delay, reverb, chorus; sixteen more effects in the online store; (iOS 7.0 or later).

Swoopster (https://holdernessmedia.zendesk.com/hc/en-us/requests/new) Flanger, fuzz, and vibrato effect designed for Audiobus and Inter-App Audio; process synths, drum machines, and hundreds of other apps by connecting through Audiobus; (iOS 7.0 or later).

ToneStack (http://www.yonac.com/tonestack/index.html) Signal processing; variety of amps and FX; latest advancements in modeling technology; (iOS 7.0 or later).

Turnado (http://sugar-bytes.de/content/products/TurnadoIOS/index .php?lang=en) Multi-effect tool crafted for real-time audio manipulation; combines unique and dynamic effects with intuitive and immediate control; MIDI Learn; Virtual MIDI port; Audiobus supported input, effects; Apple Core MIDI compliant; (iOS 7.0 or later).

UltraPhaser (http://www.elephantcandy.com/app/ultraphaser/) Wide range of phasing effects can be applied to all kinds of input sounds, from guitar, bass, and keyboard to full mixes and electronic setups; Inter-App Audio and Audiobus compatibility; (iOS 6.0 or later).

Vio (http://weareboon.com) Vocal processing; Audiobus supported input; (iOS 7.0 or later).

VocaLive (http://www.ikmultimedia.com/products/cat-view.php?C=family-vocalive) App provides vocalists with a suite of twelve real-time professional vocal effects along with a recorder for studio-quality sound in a portable package; Audiobus compatible; (iOS 6.0 or later).

WOW Filterbox (http://www.sugar-bytes.de/content/products/ WOW2IOS/index.php?lang=en) Twenty-one filter types with vowel mode, distortions; Audiobus supported input, effects; Apple Core MIDI compliant; MIDI Clock Sync; (iOS 7.0 or later).

SYNTHESIZER APPS

Addictive Synth (http://www.virsyn.net/mobileapp/index.php?option=com_ content&view=article&id=5&Itemid=8) Dynamic wavetable synthesizer; six dynamic wavetable oscillators per voice, up to forty-eight total; continuous morphing between two oscillator sets; real-time editing of up to 128 partials/overtones; real-time editing of filter structure to create arbitrary body resonances; extensive modulation possibilities using up to four LFOs and four Envelopes; control matrix allows real-time control of five parameters using the X/Y touch pad, the modulation wheel, and the tilt sensors of the iPad; monophonic or polyphonic with eight voices; 128 factory presets; unlimited user presets can be shared; up to three effects concurrently usable selected from: equalizer, phaser, flanger, chorus, and stereo/cross delay; play melodies live with the on-screen keyboard; drag fingers for slides and vibrato; velocity-sensitive keyboard; select from dozens of scales and play them with scale-optimized keyboard layout; full featured programmable arpeggiator; Audiobus compatible output; Inter-App Audio compatible instrument/generator; export loops as audio and MIDI files; audio pasteboard for exchange with other apps; exchange user presets with file sharing in iTunes; create sounds and loops; dynamic wavetable synthesis; Apple Core MIDI compliant; Virtual MIDI port; (iOS 5.0 or later).

Alchemy Synth Mobile Studio (http://www.camelaudio.com/Alchemy Mobile.php) Synthesizer studio; includes a palette of musical instruments using the remix pad to create expressive sounds; mobile version of Camel Audio's award-winning Alchemy synth, which can be heard on thousands of records by world-class musicians; powerful, expressive synthesizer; synths, basses, piano, strings, guitar, drum kits, and more; high-quality sound; large variety of sounds; intuitive song sketch pad; four-track sequencer with mixer and solo/mute; drum pads to create beats; eight songs included for remixing tracks; remix pad; morph between synth variations; tilt and inertia; sixteen performance controls; velocity sensitivity and per note filter sweeps; keyboard scales and one touch chords; external MIDI keyboard support; integrate with other apps and mobile studio; expandable with a wide range of instrument sound libraries; thirty sounds included with thirty-five more free when registered; ten in-app purchase sound libraries available including Dubstep, Ambient, and Dream Voices; each library contains fifty sounds; four hundred variations; Audiobus supported input; supports AudioCopy and General Pasteboard; supports iOS Inter-App Audio; Apple Core MIDI compliant; (iOS 5.0 or later).

AnaddrSynth (http://anaddrsynth.wordpress.com) Sound synthesis tool composed of three analog-modeling oscillator engines, three configurable oscillator banks, hi/lo/moving filters, a wah-wah effect, an echo/delay effect, iCloud patch/settings sync, and configurable attack-decay-sustain-release and modulation modules; (iOS 7.1 or later).

Animoog (http://www.moogmusic.com/products/apps/animoog-0) Powered by Moog's Anisotropic Synth Engine (ASE); Moog Music's first professional synthesizer for iOS devices; move dynamically through an X/Y space of unique timbres to create a constantly evolving and expressive soundscape; Timbre; Polyphonic Modulation; Polyphonic Pitch Shifting; Delay Module; Thick Module; Moog Filter; Record Module; Path Module; Orbit Module;

Flexible Modulation Matrix; MIDI in; Audiobus supported input; Apple Core MIDI compliant; MIDI Clock Sync; Virtual MIDI port; (iOS 5.0 or later).

Arctic Keys (http://www.onereddog.com.au/arctickeys.html) Virtual analog synthesizer; classic dual-oscillator subtractive synthesis design; professional instrument with high audio quality; MIDI Learn; Virtual MIDI port; Audiobus supported input; supports AudioCopy and General Pasteboard; Apple Core MIDI compliant; (iOS 5.0 or later).

Arctic ProSynth (http://www.onereddog.com.au/arcticpro.html) Modern subtractive synthesizer; does not aim to accurately model or re-create vintage synths of yesteryear; includes two HyperOscillators, dual filters, and master effects; Vocoder; (iOS 5.1.1 or later).

ARGON Synth (http://ios.icegear.net/argon/) Monophonic virtual analog synthesizer; stream live audio directly to other Audiobus-compatible apps; 3xOSCs; FM (XMOD); Ring Modulation; Wave Shape Modulation; Formant Oscillator; Pitched Noise Oscillator; Oscillator Sync; Filter: LP24, LP18, LP12, LP6, BP, HP; Overdrive; Three Band EQ; 3xLFOs; 3xENV (ADSR)s; Stereo Delay; Step Sequencer Arpeggiator; Loop Recorder; CoreMIDI Input; 128 Factory Presets; support for Audiobus; support for Virtual MIDI-IN and Background Audio; support for four-inch display; Apple Core MIDI compliant; (iOS 6.1 or later).

Arturia iMini (http://www.arturia.com/products/ipad-synths/imini/overview) Re-creation of the classic 1971 Minimoog synthesizer, one of the most iconic synthesizers of all time; more than five hundred sounds by leading sound designers; supports iOS Inter-App Audio; Virtual MIDI port; Audiobus supported input slot; Apple Core MIDI compliant; (iOS 5.1 or later).

Arturia iSem (http://www.arturia.com/products/ipad-synths/isem/overview) Re-creation of the classic 1974 Oberheim SEM (Synthesizer Expander Module), one of the world's first self-contained synthesizer modules; Audiobus supported input; Virtual MIDI port; supports iOS Inter-App Audio; Apple Core MIDI compliant; (iOS 5.1.1 or later).

Audulus (http://audulus.com) Build synthesizers, design new sounds, and process audio; low-latency real-time processing suitable for live performance; user interface is easy to learn; Audiobus supported input, effects; Apple Core MIDI compliant; (iOS 7.0 or later).

BassLine (http://www.finger-pro.com/bassline.html) Virtual analog bass synthesizer with built-in step sequencer and on-board effects; low-pass filter produces high-resonance filter sweeps typical of vintage synths like the legendary TB-303, the trademark of Acid House that influenced many different styles of electronic music; supports AudioCopy and General Pasteboard; Audiobus supported input; Apple Core MIDI compliant; MIDI Clock Sync; Virtual MIDI port; (iOS 4.2 or later).

Bebot—Robot Synth (http://www.normalware.com) Musical synthesizer with a multi-touch control method built into an animated cartoon robot; demo videos at website; four different synthesis modes; analog-sounding filters; effects and overdrive distortion; polyphonic synthesizer offering a wide range of sounds and a high degree of control via the multi-touch screen; Audiobus supported input; (iOS 4.3 or later).

BitWiz Audio Synth (http://kymatica.com/Software/BitWiz) Real-time BYTEBEAT synthesizer; Universal app; landscape and portrait; instant update of audio for live-coding; multi-touch XY-pad for tweaking/playing in real time; Core MIDI support for external control; record to file; export or audio-copy into other apps; stream the output to other Audiobus compatible apps; upload recordings to SoundCloud; share codes through Twitter or e-mail or open in other apps; supports sample rates 44100, 22050, 14700, 11025, 8820, etc.; stereo audio; optional microphone input for distortion effects; custom keyboard with arrow-keys for easy editing; background audio operation; many factory presets; transfer codes and recordings through iTunes File Sharing; swipe with two fingers to generate code randomly; translate C-like code expressions into eight-bit generative stereo audio in real time; explore the algorithmic music of simple bitwise arithmetic operations while watching the retro-digital 3-D graphics; Audiobus supported input, effects, output; Apple Core MIDI compliant; (iOS 5.1 or later).

CASSINI Synth (http://ios.icegear.net/cassini_ipad/) Polyphonic synthesizer with three OSCs, two filters, AMP, nine EGs, six LFOs, three-band EQ, saturator, two delays, and arpeggiator; recorder; AudioCopy; Virtual MIDI port; Audiobus supported input; Apple Core MIDI compliant; (iOS 6.1 or later).

Clap Box (http://puremagnetik.com/about) Emulation of the classic 1982 Clap Trap; microphone triggering and bpm matching; Apple Core MIDI compliant; (iOS 4.3 or later).

Crystal Synth XT (http://www.greenoak.com/crystal/Crystal/Crystal.html) Full featured semi-modular synthesizer; features MIDI keyboard input via the iPad's camera adapter or any Core MIDI-compatible MIDI interface such as the Line 6 MIDI Mobilizer and sharing of sounds with the desktop version of Crystal; MIDI keyboard input via the iPad's camera adapter or any Core MIDI-compatible MIDI interface; Audiobus supported input; Apple Core MIDI compliant; Virtual MIDI port; (iOS 5.1 or later).

Different Drummer (http://technemedia.com/drummer/) Music synthesizer; eight sampler tracks with five-wave Cyclophone Controls; Audiobus and Inter-App Audio out plus background play; stereo audio recording; export via Dropbox or iTunes or AudioCopy or General Pasteboard; Progression Composer to create guide tracks or multi-track arpeggios; control drum pitches, rests, ties, dynamics, and panning using complex waves; Instant Capture; save Wave Presets to Wave Library and share; string Wave Sets together in saved Sequences; export Wave Sets or Sequences as MIDI Files via iTunes, e-mail, or Dropbox; more than 190 drum and sound FX samples; add own samples via iTunes, AudioPaste, or Dropbox; more than seventy musical scales to assign drums to as well as key; channel volume, filter, distortion, and reverb mixer; set key, tempo, swing, and time signature; Automation for randomizing and morphing playback; MIDI Sync Out; up to eight channels of MIDI Out; Core MIDI lets user play other synthesizers and drum machines on iPad; Audiobus supported input; supports AudioCopy and General Pasteboard; MIDI Clock Sync; (iOS 6.1 or later).

DXi FM Synthesizer (http://www.taktech.org/takm/DXie/DXi_for_iPhone.html) Inspired by the popular 1980s FM synthesizer; four-operator FM synth; sixteen-step loop sequencer; includes sixty-nine preset sounds and twenty-nine blank preset spaces to create own sounds; eight sine-wave variations; effects section; video introduction on website; Virtual MIDI port; Audiobus supported input; (iOS 7.0 or later).

EGSY01 Analog Synth (http://www.elliottgarage.com/software/) Inter-App Audio, Core MIDI, Audiobus; create sounds from modern to vintage analog; work with arpeggiator, step sequencer, and FM; addictive and subtractive synthesis; (iOS 5.1 or later).

Electric Piano Synthesizer and ***Electronic Piano Synthesizer XS*** (http://www.bacaj-apps.com/Electronic_Piano_Synthesizer.html) App is based on a thirty-two-bit real-time sound engine; sounds are computed; no samples used; supports AudioCopy and General Pasteboard; Audiobus supported input; Apple Core MIDI compliant; MIDI Learn; (iOS 5.1 or later).

Epic Synth (http://www.epicsynth.com) Polyphonic analog-like synthesizer inspired by classic synths of the 1980s; Audiobus supported input; Apple Core MIDI compliant; (iOS 6.0 or later).

Geo Synthesizer (http://www.wizdommusic.com/products/geo_synthesizer.html) Musical instrument and MIDI controller for a multi-touch surface; Virtual MIDI port; Audiobus supported input; Apple Core MIDI compliant; Universal app; comprehensive manual available at website; (iOS 5.0 or later).

Grain Science (http://www.wooji-juice.com/products/grain-science/) Synthesizer designed for musicians, soundscape artists, and SFX engineers; built on principles of granular synthesis; mixes it up with traditional synthesis techniques to make a hybrid synthesizer; Audiobus supported input; Apple Core MIDI compliant; MIDI Learn; Virtual MIDI port; (iOS 6.0 or later).

GrainBender (http://www.plastaq.com/grainbender/) Synthesizer for designing new sounds; tap the bend button and sculpt a new sound directly in real time; capable of making a wide range of sounds, from traditional analog favorites to exotic digital effects; Audiobus supported input; supports AudioCopy and General Pasteboard; (iOS 7.0 or later).

GRIT (http://twisted-electrons.com/apps/grit/) Bass synthesizer with real-time effects and filters; fifty patches with professionally crafted samples; seven expression knobs that have a different effect depending on preset: Cutoff, Resonance, Lfo Speed, Lfo Depth, Bit Crush, Attack, Release; Audiobus; MIDI input; demos available on website; (iOS 6.1 or later).

GyroSynth (http://beepstreet.com/gyrosynth) Gesture-driven music synthesizer that takes full advantage of the gyroscope; play and modulate the sound by moving hand through the air like the legendary theremin instrument; makes use of the gyroscope to measure true roll, pitch, and yaw and translates them to sound parameters like pitch, volume, modulation, or filter cutoff; play melodies perfectly in tune; synthesizers: vocal (formant synthesizer), sine wave, square (PWM modulation), saw, granular sample player with recording function and R2D2; experimental FM modulation; effects: delay and boost; several control modes; included musical scales allow user to play in several styles; built-in recorder and player with overdub function; Audiobus supported input; (iOS 5.0 or later).

Impaktor (http://beepstreet.com/impaktor) Drum synthesizer with a large sonic palette; turns any surface into a playable percussion instrument; Audiobus supported input; (iOS 5.0 or later).

iSyn Poly (http://www.virsyn.net/mobileapp/index.php?option=com_content&view=article&id=2&Itemid=8) Electronic music studio with three studio-quality, fully programmable virtual analog synthesizers and a drum machine; (iOS 6.1 or later).

iSynthesizer (http://amitech.co) Classic monophonic synthesizer; designed for live performance; fast switchable user interface designed with the help of professional artists; generate high-quality sounds; quickly switch between presets live; monophonic playing one note at a time like a classic synthesizer; configurable presets; save favorite settings; oscillator module; dual oscillators;

square, triangle, sawtooth, reverse sawtooth waves; octave selector, supporting five octaves; glide between notes (lag processor); oscillator fine-tune more than one full octave; oscillator sync; low-pass filter module; modulation module; tremolo/wave; vibrato/pitch; filter; envelope generator module; supports ADSR (attack, decay, sustain, release); volume; filter (amount of low-pass filter); arpeggio; (iOS 4.0 or later).

iVCS3 (http://www.densitygs.com) Official EMS VCS3 emulator; first portable commercially available synthesizer; (iOS 6.1 or later).

iVoxel (http://www.virsyn.net/mobileapp/index.php?option=com_content &view=article&id=1&Itemid=2) Combination of a voice-optimized synthesizer and a vocoder; based on the Matrix vocoder from VirSyn used by many famous artists including Kraftwerk; Virtual MIDI port; Audiobus supported input, effects; Apple Core MIDI compliant; (iOS 5.0 or later).

Jasuto Pro (http://www.jasuto.com/main/) Modular synthesizer; visually construct synths/effects and make sequences on device; sampler to record/ resample/edit; create new samples by drawing in the app; supports AudioCopy and General Pasteboard; (iOS 5.1 or later; Android 1.5 and up).

KORG iMS-20 (http://www.korg.com/us/products/software/ims_20_for_ ipad/) Analog synth studio; complete re-creation of the Korg MS-20 synth, an analog sequencer, a drum machine, and Korg's Kaoss Pad technology; share songs online via SoundCloud; supports AudioCopy and General Pasteboard; Audiobus supported input; Virtual MIDI port; Apple Core MIDI compliant; (iOS 5.1 or later).

KORG iPolysix (http://www.korg.com/us/products/software/ipolysix_for_ ipad/) Analog polyphonic synthesizer; sequencer, drum machine, and mixer; transforms iPad into an analog synth studio; supports AudioCopy and General Pasteboard; Audiobus supported input; Apple Core MIDI compliant; Virtual MIDI port; (iOS 5.1 or later).

Magellan (http://www.yonac.com/magellan/) Professional analog modeling synthesizer; two independent polyphonic synth engines to be used individually or together; full FX rack; six oscillators; dual filters per engine; multiple unison stages; extensive modulation matrix; dedicated arps; polyphonic step-sequencer; MIDI Clock Sync; supports iOS Inter-App Audio; MIDI Learn; Virtual MIDI port; supports AudioCopy and General Pasteboard; Audiobus supported input, effects; Apple Core MIDI compliant; (iOS 5.1 or later).

Mellotronics M3000 (http://www.omenie.com) App is an authentic replication of the legendary M400 tape replay instrument with many additional features; (iOS 5.1 or later).

microTERA (http://www.virsyn.net/mobileapp/index.php?option=com_co ntent&view=article&id=14&Itemid=8) Waveshaping synthesis is a type of distortion synthesis that can create dynamic spectra in a controlled way; in waveshaping, it is possible to change the spectrum with the amplitude of the sound such as the clipping caused by overdriving an audio amplifier; (iOS 6.0 or later).

MIDI Guitar (http://jamorigin.com/products/MIDI-guitar/) Turn any guitar into a polyphonic guitar-synthesizer or record tablature by playing it; Virtual MIDI port; (iOS 4.3 or later).

MiniMapper (http://wolfgangpalm.com/ios-products/minimapper/) Performance synthesizer with a wide range of sounds; Audiobus supported input slot; Apple Core MIDI compliant; Virtual MIDI port; (iOS 5.1 or later).

miniSynth PRO (http://www.yonac.com/software/miniSynthPRO/) App is a full featured, professional-grade virtual analog synthesizer; Apple Core MIDI compliant; (iOS 4.2 or later).

Mitosynth (http://www.wooji-juice.com/products/mitosynth/) Hybrid synthesizer with one hundred built-in patches; add rich modulation, effects, and filters; Core MIDI; Virtual MIDI; Audiobus support; AudioCopy/Paste; AudioShare; Dropbox; Universal app; (iOS 7.0 or later).

Modular Synthesizer (http://www.pulsecodeinc.com/index.html) Sound design tool; build a custom synthesizer by connecting different sound modules together; Audiobus supported input; Apple Core MIDI compliant; (iOS 6.0 or later).

Nanologue (http://www.steinberg.net/en/products/mobile_apps/nanologue .html) Sounds and sound effects; multi-touch interface; monophonic synthesizer; used stand-alone and with iOS host application via Inter-App Audio; brings the power of VST 3 technology to the iOS platform; (iOS 7.0 or later).

Nave (http://www.waldorf-music.info/nave-overview) Waldorf's first synthesizer app; sound engine includes two novel wavetable oscillators; spectrum of a sound can be transposed independently of its pitch; sounds with an accent on formants can be reproduced; integrated speech synthesizer for the creation of wavetables; supports AudioCopy and General Pasteboard; Audiobus supported input; Apple Core MIDI compliant; MIDI Clock Sync; MIDI Learn; Virtual MIDI port; supports iOS Inter-App Audio; (iOS 6.0 or later).

Nlog MIDI Synth (http://www.temporubato.com) App is a professional virtual analog synthesizer; supports Audiobus, Inter-App Audio, WIST, and Core MIDI compatible apps and interfaces; (iOS 5.1.1 or later).

NlogSynth PRO (http://www.temporubato.com/index.php?page=ProdPRO)
Professional virtual analog synthesizer supporting Core MIDI, WIST, and interfaces from Akai, Alesis, Line6, IK Multimedia, and others; MIDI Learn;
supports AudioCopy and General Pasteboard; Audiobus supported input, effects, output; Apple Core MIDI compliant; supports iOS Inter-App Audio;
Virtual MIDI port; (iOS 5.1.1 or later).

Novation Launchkey (http://us.novationmusic.com/software/launchkey-app)
Synthesizer for iPad; includes eighty synth sounds for performing and producing music; based on Novation synthesizer technology; Audiobus supported
input; (iOS 5.0 or later).

Phawuo (http://alexnadzharov.ru/phawuo/) App is a monophonic virtual-
analog synth with "dirty" oscillators; good for bass sounds and effects; (iOS
7.0 or later).

PixelWave (http://www.warmplace.ru/soft/synths/) Experimental synth
with old-school pixel interface; Audiobus supported input; JACK Audio/
MIDI Connection Kit supported app; (iOS 4.3 or later; Android 2.3 and up).

PPG WaveGenerator (http://wolfgangpalm.com/ios-products/wavegenera
tor/) From the inventor of wavetable synthesis, Wolfgang Palm; next-
generation synthesizer building on the heritage of the PPG Wave keyboards;
Audiobus supported input; Apple Core MIDI compliant; Virtual MIDI port;
(iOS 6.0 or later).

PPG WaveMapper (http://wolfgangpalm.com/ios-products/wavemapper/)
Second synthesizer to complement the PPG WaveGenerator from Wolfgang
Palm; create limitless sounds; Virtual MIDI port; Apple Core MIDI compli-
ant; Audiobus supported input; (iOS 6.0 or later).

Pro Keys (http://beepstreet.com/prokeys) Multi-instrumental polyphonic
keyboard with sounds of legendary synthesizers; playable interface with two
independent piano/drum keyboard modules and seventeen studio quality
presets; overdubbing, loop player, and vocal recorder; mirrored keyboard dual
mode; twenty presets; fourteen instruments and six drum kits; independent
loop player and recorder with WiFi file sharing; record complex songs track
by track; vocal recorder requires headphones; two independent keyboard
modules; ultra-low latency; high-quality thirty-two-bit audio processing;
drum pads with pitch controller; adjustable keyboard size with full-size keyboard or two octaves; realistic pitch bend; poly and legato modes; full-screen
mode; multitap delay FX; MIDI input split keyboard option supports MIDI
Mobilizer and Core MIDI; Apple Core MIDI compliant; Virtual MIDI port;
(iOS 4.3 or later).

RD4—Groovebox (http://www.mikrosonic.com/rd4) Music-making app with virtual analog synthesizers, drum machines, and effects; (iOS 7.0 or later; Android 4.0.3 and up).

Reactable Mobile (http://reactable.com/products/mobile/) Create and improvise music in an intuitive and visual way; power of Reactable instrument; (iOS 5.1 or later; Android 2.2 and up).

ReBirth for iPad (http://www.rebirthapp.com) Propellerhead Software's legendary Techno Micro Composer resurrected and customized for the iPad; emulates dance music's three backbone devices: the Roland TB-303 Bass synth and the Roland TR-808 and 909 drum machines; FX units; pattern sequencers; built-in sharing features; Audiobus supported input, slot; supports AudioCopy and General Pasteboard; MIDI Clock Sync; (iOS 5.0 or later).

Rhythm Studio (http://www.pulsecodeinc.com/rhythmStudio.html) Universal electronic music-making app; re-creations of classic synths and drum machines and other studio hardware; Audiobus supported input; supports AudioCopy and General Pasteboard; (iOS 4.3 or later).

shapesynth (http://www.humbletune.com/shapesynth/) Polyphonic synthesizer allowing the user to draw the shape of the oscillator waveform; Audiobus supported input; Apple Core MIDI compliant; (iOS 4.3 or later).

SIDPAD (http://twisted-electrons.com/apps/sidpad/) Three voice synthesizer; more than forty controllable parameters to create a wide range of eight-bit chip-tune sounds; all knobs MIDI controllable; Audiobus supported input; (iOS 6.1 or later).

Slide Control Fluid Synth (http://www.rouetproduction.com) Music application for performing and recording musicians; 890 soundfont sounds; choose any scale and slide finger to play any melody; Virtual MIDI port; (iOS 4.2 or later).

SoundFont Pro (http://www.itimsystems.com/product-soundfontpro/) Soundfont player with many features; record music and vocals and share; can overdub repeatedly in a non-destructive manner and build composition over time; suitable for studio use as well as for live gigs; multi-timbral, multi-layered, and multi-zonal; (iOS 7.0 or later).

SpectrumGen (http://www.warmplace.ru/soft/synths/#sg) Spectral synthesizer with a simple and effective user interface; multi-touch arpeggiator; export to WAV via iTunes File Sharing; copy to share with other iOS sound apps; eight predefined timbres; changeable number of octaves from two to eight; draw the spectrum and play it at the same time; MIDI-IN support; Audiobus and JACK support; (iOS 4.3 or later).

SquareSynth (http://jnapps.wordpress.com/square/) Table-based synthesizer based on old-school chip-tune trackers; create sounds similar to eight-bit game consoles and computers; Apple Core MIDI compliant; (iOS 5.0 or later).

Stria (http://www.densitygs.com) App is a multilevel interactive sound synthesizer; up to 240 frequency modulation oscillators or simple wavetable additive; AudioCopy and AudioPaste; Dropbox support; Audiobus and Inter-App Audio compatible; (iOS 6.1 or later).

Sunrizer Synth (http://beepstreet.com/sunrizer) Virtual analog synthesizer; works with any MIDI keyboard or sequencer; SuperSaw sound emulation; oscillators; multifilters; effects; supports AudioCopy and General Pasteboard; Audiobus supported input, slot; Apple Core MIDI compliant; MIDI Clock Sync; MIDI Learn; Virtual MIDI port; (iOS 5.0 or later).

SunVox (http://www.warmplace.ru/soft/sunvox/) Modular synthesizer with pattern-based sequencer; Audiobus supported input, effects; Apple Core MIDI compliant; Virtual MIDI port; JACK Audio/MIDI Connection Kit supported app; (iOS 4.3 or later; Android 2.3 and up).

Super Jupiter 8V Programmer (http://www.4littlefonzies.nl) For programming the Arturia Jupiter 8V virtual instrument; Apple Core MIDI compliant; (iOS 4.3 or later).

Sylo Synth (http://www.wooji-juice.com/products/sylo-synth/) Allows user to play music or create sound effects; includes a collection of pre-designed sounds to use or tweak or make new sounds from scratch; supports AudioCopy and General Pasteboard; (iOS 6.0 or later).

Synth (http://www.retronyms.com/synth/) Polyphonic synthesizer; more than forty instruments; mod wheel; pitch-bend wheel; adjustable delay and distortion; sampler included; (iOS 3.2 or later).

Synth Arp and Drum (http://usa.yamaha.com/products/apps/synth_arp/) Create phrases and beats; compatible with the Inter-App Audio function (iOS 7 and above); Audiobus support, input; perform an audio mixdown of recorded songs; Audio library compatible with SoundCloud and AudioCopy; MIDI channel is displayed when initializing the DrumPad EDIT settings; additional nineteen tones and twenty-four arpeggiator patterns; arpeggiator and drum pad that allows user to play the internal synthesizer or any connected MIDI device and produce music with phrases in a variety of musical styles; arpeggiator automatically plays the individual notes of a chord in a selected pattern; 342 arpeggiator patterns can play phrases from all kinds of music genres with one finger; (iOS 6.1 or later).

Synthecaster (http://synthecaster.wordpress.com) Fuses elements from guitars and keyboards; full-range polyphonic instrument; accommodates many playing styles and generates a wide variety of sounds; allows for technical playing styles such as shredding and complex chord formation; can be used as a MIDI controller; motion modulation; (iOS 7.0 or later).

synthmate (http://soh.la/synthmate.html) Polyphonic synth for iPad; real-time control, low latency; multi-touch interface, change anything while playing; five note polyphony; Audiobus supported input; Apple Core MIDI compliant; (iOS 5.1 or later).

SynthTronica (http://synthtronica.com) App is a programmable polyphonic spectral synthesizer; supports AudioCopy and General Pasteboard; (iOS 3.2 or later).

SynthX (http://www.wayoutware.com/synthx/) Virtual analog synthesizer for the iOS platform; multi-touch user interface; MIDI Beat Clock LFO sync; supports AudioCopy and General Pasteboard; Apple Core MIDI compliant; (iOS 4.2 or later).

TANSU Synth (http://tansusynth.music-airport.co.jp/en/) Miniature version of a modular analog synth; same one used and mastered by Hideki Matsutake of Logic System; (iOS 5.0 or later).

TC-11 (http://www.bitshapesoftware.com/instruments/tc-11/) Programmable multi-touch synthesizer for the iPad; choose from 120 included presets or build own using touch controls, device motion, and on-board modules; supports AudioCopy and General Pasteboard; Audiobus supported input; (iOS 6.1 or later).

Tf7 Synth (http://tenaciousfrog.com/tf7-synth/) FM synthesizer; play music by playing shapes across the interface; includes twenty sounds; morph sounds while playing; (iOS 6.1 or later).

the microTone (http://the-microtone.appsios.net) Microtonal polyphonic synthesizer and MIDI/OSC controller; Apple Core MIDI compliant; (iOS 5.0 or later).

Thor Polysonic Synthesizer (https://www.propellerheads.se/products/thor/) Reason's legendary flagship synth; sound sculpting capabilities; innovative keyboard design; expressive instrument for the iPad; Audiobus supported input; Apple Core MIDI compliant; (iOS 5.1 or later).

Uber Synth (http://intelligentgadgets.us/apps.shtml#2) Polyphonic multi-timbral FM synthesizer that runs on any iOS device; Audiobus supported input; (iOS 5.1 or later).

Unity Synthesizer (Steve O'Connell) Create synth sounds with on-screen or external MIDI keyboard; base configuration includes two synth engines: Retro AS-1 analog and Unity DS-1 sample playback; twenty-three audio effects; seven sound banks including General MIDI, Best of Retro AS-1, Pipe Organ, Rock, and Pop and drum and percussion loops banks; arpeggiator, sequencer, and scales can be used for real-time MIDI playback; Virtual MIDI port; Audiobus supported input; Apple Core MIDI compliant; (iOS 7.1 or later).

Virtual Synth (http://www.riteshlala.net/home/virtual-synth-for-iphone/) Experimental synthesizer; generate sine tones and square waves from a wide range of frequencies; experiment with different grid positions to generate modulating signals with different beat frequencies; minimal control interface; touch to create an instrument anywhere on the grid; numerical labels display frequency values; select a different instrument/frequency and tap on an existing instrument to change it; beat frequency for modulating waves increases diagonally; tap with two fingers at any time to remove all instruments; (iOS 3.0 or later).

Voice Synth (http://www.voicesynth.com) Professional live instrument to create new voices, choirs, sounds, and soundscapes based on user's own unique voice; includes Inter-App Audio, three twenty-four-band live vocoders, AutoPitch, multiple voice harmony arranger, spectrum stroboscope, pitch and formant shifter, sampler, twenty-four-band equalizer, distortion, delay, chorus, and reverb; Audiobus supported input, effects, output; (iOS 7.0 or later).

XENON Groove Synthesizer (http://ios.icegear.net/xenon/) Polyphonic hybrid synthesizer; two x VA monophonic synthesizer, polyphonic PCM synthesizer, rhythm machine, sequencer, and mixer; includes more than 350 preset sounds; Apple Core MIDI compliant; (iOS 6.1 or later).

xMod (http://intelligentgadgets.us/apps.shtml#2) Synth designed for iOS devices; sound engine is a polyphonic triple-oscillator cross-modulated FM synthesizer; FM is widely regarded as the most efficient way of generating complex organic musical sounds; Audiobus supported input; Apple Core MIDI compliant; (iOS 5.1 or later).

Z3TA+ (http://blog.cakewalk.com/z3ta-on-ipad/) Legendary synthesizer used on countless records; hundreds of presets, modulatable waveshaping, dual mode filters, and flexible effects; on-screen keyboard and external MIDI control; Inter-App Audio, Background Audio, and Audiobus support; (iOS 7.0 or later).

zMors (http://www.zmors.de) Four-layer synthesizer with built-in sequencer; (iOS 7.0 or later)

• *4* •

Recording Music: "The Gear" Apps

DJ APPS

Baby Scratch (http://www.async-games.com/baby.html) DJ turntable with a built-in sampler; scratch own voice; Flare Scratch engine; battle record with classic DJ samples; three built-in beat loops; level fader; transformer button; (iOS 4.0 or later).

Cloud DJ (http://www.cloud-dj.net/en/) Stream music from the Internet and mix it like a DJ; songs updated daily to SoundCloud form the basis of DJ library; choose tracks and play back immediately; tap waveforms to scratch; automatic bpm detection and tempo matching; pre-cueing feature; use up to six samplers; time stretch plus four other effects included; Audiobus supported input; (iOS 5.1 or later).

CTRL Music (http://www.8linq.com) Music-remixing app; use music from iPod library or stream from SoundCloud; more than twenty DJ-style effects and synths to use and master; effects control via touch and tilt; effects are in time with the music and synths are in key; trigger multiple effects simultaneously; Universal app; (iOS 6.0 or later).

DJ Control (http://trajkovski.net) Wireless DJ MIDI controller; exact emulation of Hercules's DJ Console RMX DJ MIDI controller functions by using MIDI over WiFi; multi-touch controls; DJ seek/scratch wheels; pitch/tempo sliders; channel level sliders; master volume control; crossfader slider; cue and play buttons; cue points buttons; beat lock, beat sync buttons; PFL/cue select buttons; effects on/off button; complete EQ with gain; song list navigation pane; MIDI over WiFi connection; sixteen case materials; twelve jog wheel plates; two knob styles; tablet/pad support; hardware acceleration; (Android 2.1 and up).

Djay 2 (http://www.algoriddim.com/djay-ipad) DJ app; iTunes music library integration; Spotify integration; Automix; Audio FX: Flanger, Phaser, Echo, Gate, Bit Crusher; FX Expansion Packs powered by Sugar Bytes; mixer, tempo, pitch-bend, filter, and EQ controls; looping and cue points; colored HD waveforms; live recording (iTunes only); sampler with included sound packs by Snoop Dogg, DJ QBert, Milk & Sugar, and more; single deck mode; pre-cueing with headphones; advanced time-stretching key lock; key detection and matching (requires iPad); automatic beat detection; auto-gain; iCloud integration to sync metadata; iTunes Store integration; support for Inter-App Audio, Audiobus, and all major audio formats; support for AirPlay and Bluetooth devices; support for DJ MIDI Controllers: Pioneer DDJ-WeGO, DDJ-WeGO2, DDJ-ERGO, Vestax Spin 2, Numark iDJ Pro, Mixdeck Quad, iDJ

Live, iDJ Live II, ION iDJ 2 Go, Reloop Beatpad, Casio XW-J1, and Philips M1X-DJ; (iOS 7.0 or later).

edjing—DJ Music Mixer Studio (http://www.edjing.com) Sharing function to share mix on Facebook, Twitter, and Google+; access music library to mix tracks; access to SoundCloud 2 Turntables; Cross Fader; free DJ effects (FX) include Cue Point, Flanger, Equalizer, Scratch/Pitch on Vinyl; gyro FX option; wide sound spectrum for optimum beats localization; bpm synchronization; recording in HD CD quality, wave format; AutoMix mode for automatic transitions between titles of playlists; remix MP3 files of more than one hour; additional effects (FX) for purchase include: Pre-Cueing, Reverse, Double Flipping, Auto Scratch, Reverbs, Gate, TK Filter; compatible with Airplay; (iOS 6.0 or later; Android, varies with device).

iDJ 2 GO (http://www.ionaudio.com/products/details/idj2go) Create DJ mixes with iTunes music library; DJ controls on touch screen; two turntables to scratch and mix; crossfader; share mixes; sync control automatically matches beats; pitch controls and three-band EQs on each deck; interactive tutorial; works best with the iDJ 2 GO controller; (iOS 6.0 or later).

Micro DJ (http://psychobearstudios.com) Turn music into new creations; audio editing device; edit the pitch, speed, and tempo of any song; create interesting sound effects; (iOS 3.2 or later).

Mixcloud—Radio & DJ Mixes (http://www.mixcloud.com/mobile/) Free radio shows, DJ mixes, and podcasts; search, browse, discover, and stream a large collection of high-quality audio; follow favorite DJs and presenters; (iOS 7.0 or later; Android 2.3 and up).

Tap DJ—Mix and Scratch Your Music (http://tap.dj) Pocket DJ app for iPhone and iPod; scratch, mix, and add FX to iPod music; (iOS 4.3 or later).

Touch DJ Evolution—Visual Mixing, Key Lock, AutoSync (http://aMIDIo.com/dj/) App does not re-create any existing hardware DJ setup like turntables or CD platter; offers a fresh approach, giving the ability of direct track manipulation; (iOS 4.2 or later).

DIGITAL AUDIO WORKSTATIONS (DAWS),
SEQUENCERS, SAMPLERS, AND RECORDING APPS

Air Recorder (http://www.roland.com/apps/) Record sound through internal mic of iPhone/iPod Touch or wirelessly make digital recordings of the sound from a Roland electronic musical instrument; load songs from music library of iPhone and play them back while playing and recording instrument; supports AudioCopy and General Pasteboard; (iOS 5.1 or later).

Analyzer (http://dspmobile.de/2012/10/analyzer-2-0/) Combination of a sound pressure level (SPL) meter and a full-range multiple-bands frequency analyzer; designed for audio professionals who need to evaluate different working environments such as studios and live stages; use for measuring stages, testing speakers, setting up a HiFi system, evaluating sound-mixing environments, or analyzing an instrument; intuitive user interface with zoom gesture control, individual color settings, and zoom; can store, load, and export measurement data; supports iCloud storage; Audiobus supported effects, output; (iOS 5.1 or later).

Audio Mastering (http://audio-mastering-studio.blogspot.com) Fully functional professional-quality audio-mastering application; take final mixes to the next level; precise control of all parameters and highest-quality audio processing; all-in-one tool allows user to finalize the completed mix, record a track, convert audio formats, change sample rate, convert bit depth, cut part of a track for preview, and apply fade-in and fade-outs; support for Inter-App Audio and Audiobus allows use with other music applications; external audio interface; use as a powerful sound processor or insert effect in studio and integrate with other studio hardware; tweak all the controls in real time while auditioning the result; two different control modes; adjust the final mix in basic mode or fine-tune the settings in more detail in advanced mode; includes built-in presets to find a suitable starting point, then tweak the processing to get a desired result; support for iTunes File Sharing; Inter-App Audio compatible; support Audiobus in effect or output slots; recording from external input with Inter-App Audio or Audiobus; audio-processing tool only; does not have audio editor functions; technical details, features, description, and support on website; (iOS 6.0 or later).

Audiobus (http://audiob.us) Inter-App Audio routing system; connect Audiobus-compatible music apps together like virtual cables; connect the output of one Audiobus-compatible app into the input of another; play a synthesizer live into a looper or multi-track recorder; use one app to manipulate the live output of another; multi-routing feature in-app purchase allows unlimited connections and effect chaining; presets let user save and share connections and settings of individual apps within the Audiobus workflow; multi-channel input hardware support; hardware latency controls; more than four hundred supported apps including: GarageBand, djay, Animoog for iPad, Alchemy, all KORG apps, Rebirth for iPad, Figure, Cubasis, BeatMaker2, PPG Wavegenerator and Wavemapper, DM-1, Samplr, JamUp Pro, NLog Synth Pro, Sunrizer, Loopy, SoundPrism Pro, MultiTrack DAW, Sir Sampleton, FunkBox, ThumbJam, DrumJam, Arctic Keys, Glitchbreaks, Orphion, Audulus, Magellan, Echo Pad, AudioShare, Remaster, MoDrum Rhythm Composer, Cloud DJ, NodeBeat, Auria, and Guitarism; (iOS 7.0 or later).

AudioCopy (http://retronyms.com/audiocopy/) Paste sound in hundreds of compatible apps; catalog for sounds; browse and audition all sounds copied; import own sound libraries; store sounds centrally for easy use in apps; built-in sound recorder; record, pause, record some more; sound editor; trimming, normalize, fade, and reverse; zoom in for more detail in gesture-based editor; capture audio and share directly to SoundCloud; (iOS 6.1 or later).

AudioShare (http://kymatica.com/Software/AudioShare) Sound file manager with import and export abilities; organize sound files and MIDI files on device; record, trim, convert, normalize, upload, download, export, import, zip, unzip; create folders; rename and move files and folders around; built-in Inter-App Audio host with four node slots; record Inter-App Audio compatible instruments and effects or process files through the chain; transfer files between apps and between device and computer; audiocopy sounds from music-making app and import into the AudioShare library for later sharing or copying into other apps; record live-jam from other Audiobus apps; record directly in the app or from hardware, Inter-App Audio, or other Audiobus compatible apps; preview and play sound files with waveform display and looping, trim and normalize sound files, and convert to other file formats; Inputs: record external input such as microphone directly in app; record the live output of other Audiobus compatible apps; record Inter-App Audio node apps; AudioPaste Sonoma Wireworks, General Pasteboard, Retronyms ACP2.0; import files and folders from Dropbox; download sound files from the web with built-in browser; "Open in" from other app; iTunes File Sharing; import songs from iTunes music library; Outputs: playback directly in app; AudioCopy Sonoma Wireworks, General Pasteboard, Retronyms ACP2.0; SoundCloud upload; export files and folders to Dropbox; "Open in" another app; send by e-mail; iTunes File Sharing; play back sound files into other Audiobus compatible apps; audio document manager; (iOS 5.1 or later).

Auria (http://www.auriaapp.com) Digital audio recording system; forty-eight tracks of simultaneous playback of stereo or mono files; up to twenty-four tracks of simultaneous recording when used with compatible USB audio interfaces; twenty-four-bit recording; vintage-inspired ChannelStrip on every channel by PSPAudioware includes Expander, Multiband EQ, and Compressor; MasterStrip on all subgroup and master channels featuring PSPAudioware BussPressor, EQ, and Mastering Limiter; sixty-four-bit double-precision floating point mixing engine; third-party plug-in support available via in-app purchase; MIDI Sync support with MTC Chase, MIDI Clock, and MMC; MIDI Remote Control; supports sample rates of 44.1KHz, 48KHz, and 96KHz, at twenty-four bits; AAF import and export allows transferring complete sessions between DAWs like Logic, Pro Tools, Nuendo, Samplitude, and others; convolution reverb plug-in with included IR library by MoReVoX; ClassicVerb reverb, StereoDelay, and StereoChorus plug-ins included; eight assignable subgroups and two aux sends; Time Stretching using Dirac 3 Pro technology; real-time audio scrubbing; Ripple Edit mode; Tempo sync and side chain support for plug-ins; waveform editor with cut/copy/paste, crossfade, duplicate, separate, gain, normalize, dc offset, reverse, and more; flexible snapping tools allow snapping to events, cursor, bars, beats, and more; Dropbox, Sound-

Cloud, and AudioCopy/Paste support; Inter-App Audio support; Audiobus support; track freeze for minimizing CPU usage; full automation support on all controls with graphical editing; true 100-mm faders when used in Portrait Mode; optional video import feature allows sync of video to an Auria project; Timeline ruler options include minutes:seconds, bars:beats, samples, and SMPTE time; auto-punch mode; WIST support for wireless syncing of other compatible music apps; AuriaLink allows two iPads running Auria to play and record in sync, allowing for ninety-six tracks of playback and forty-eight tracks of recording; full delay compensation on all tracks, subgroups, and aux sends; adjustable metering modes including pre- or post-fader, RMS, and peak; adjustable pan laws; sample accurate loop function; automatic sample rate conversion; metronome; (iOS 5.0 or later).

Aurora Sound Studio HD (http://www.4pocketsaudio.com/product .php?p=4) Pattern-based musical-sequencing software; allows a mixture of pattern-based recording and live performance; comprehensive help facility and video tutorials available on YouTube; fourteen instrument tracks; twelve note polyphony per track; four independent engines including Drum Machine, Analog Synth, Pad Synth, and Sampler; thirty-two patterns per song; chain up to 199 patterns in Song mode; three effect sends per track; nine effects including delay, reverb, chorus, etc.; allows creation of custom instrument patches; Layer Automation; XY Mode; splice and dice audio with Atomizer Mode; in-built sample recording; export songs as WAV and AAC; iPad expands on the iPhone version making use of the extra screen size to offer additional functionality including: full screen mixer, new effects rack view, solo and mute individual layers, new layer mix palette, synth engine features combined into single screen, file sharing via FTP, accelerometer to control XY mode, upload and share songs on the Online Song Library, MIDI export; Audiobus supported input; (iOS 6.1 or later).

Beatsurfing (http://beatsurfing.net) Organic MIDI controller builder; allows user to draw a three-dimensional controller and use it in two different and complementary ways; can tap it or surf fingers along routes, colliding with objects, triggering samples or effects; can control any MIDI-enabled device; in-app editing system; integrates seamlessly in any existing studio or live setup; Virtual MIDI port; Apple Core MIDI compliant; (iOS 6.0 or later).

bismark bs-16i (http://www.bismark.jp/bs-16i/index.html) Sixteen multi-timbral playback sampler; supports SoundFont; can be used for keyboard instrument, MIDI sound module, and MIDI player; supports all standard MIDI messages; as a keyboard instrument, can play with scalable keyboard, pitch-bend wheel, and many control change controllers; internal MIDI player supports SMF (Standard MIDI File) format; AudioCopy; SoundFont and SMF files can be sent using the file sharing functions of iTunes and "Open in" from other apps (Dropbox, etc.); supports AudioCopy and General Pasteboard; Virtual MIDI port; Audiobus supported input; Apple Core MIDI compliant; (iOS 6.1 or later).

Brainwave Sequencer (Alexandre Jean Claude) Clip-based MIDI sequencer; streamlined and intuitive MIDI clip editor; assemble and arrange songs with the arrangement editor; control up to sixteen instruments using the built-in sampler; use the built-in audio samples or use own samples; produce and export MIDI files to be imported to DAW software; Virtual MIDI port; Apple Core MIDI compliant; (iOS 4.3 or later).

ChordPolyPad—MIDI Chords Player (http://laurentcolson.com/chordpoly pad.html) For MIDI instruments, sequencers, and all other MIDI sequencers or virtual instruments installed on the same iPad; sixteen chords pads; eight groups of sixteen pads by preset (128 pads); MIDI port and MIDI channel assignable for each pad; customizable X/Y controller for each pad; velocity assignable for each chord note; strumming setup by pad; features copy and paste for pads and groups; drag chords directly from library; searching for chords available in scales; random chords for instant inspiration; multitasking; internal sound bank; receiving MIDI notes and control from external devices; send MIDI to a virtual port that can be used as MIDI input by other Core MIDI compatible applications managing multitasking on the same device; presets manager; undo/redo; supported MIDI connections: MIDI WiFi Network (RTP MIDI); Apple USB camera connection kit; any Core MIDI compatible interface; Audiobus support; the chords can be played live from pads; Apple Core MIDI compliant; (iOS 5.0 or later).

Cloud Audio Recorder (http://usa.yamaha.com/products/apps/cloud_ audio_recorder/?mode=model#page=4&mode=paging) Record musical instruments to iOS devices via the built-in microphone; recorded data can be normalized, trimmed, and signal processed; recorded data can be freely uploaded to or downloaded from SoundCloud; copy and paste sound to other iOS apps; equalizer, noise suppression, reverb, gain; undo function; supports AudioCopy and General Pasteboard; (iOS 5.0 or later).

Cotracks (http://futucraft.com/cotracks) Collaborative music studio for teamwork on a single iPad; create multiple layers of loops and phrases using multiple instruments; record and play back own parts in sync with others; multi-track sequencer; interface; XY-modulation control; editing; polyphonic synthesizer with dozens of high-quality instrument presets; sample player with a library of professional-quality samples from Inspire Audio; Audiobus support; Audiocopy support; export and import sessions via iTunes in XML format for backup and sharing; on-screen help and tips; online support integrated in app; video tutorials; (iOS 5.0 or later).

csGrain (http://boulangerlabs.com) Real-time audio processing and recording tool; create new sounds and musical textures; stereo granular sound processor; ten additional professional audio effects all realized through Csound orchestra; rendering, processing, sampling, resampling, synthesizing, resynthesizing, playing, reversing, delaying, triggering, gating, compressing, limiting, chorusing, flanging, echoing, filtering, pitch-shifting, harmonizing, granulizing, and recording in any combination all in real time; software synthesizer and signal processor; supports AudioCopy and General Pasteboard; Apple Core MIDI compliant; (iOS 7.0 or later).

Cubasis (http://www.steinberg.net/en/products/mobile_apps/cubasis.html) Steinberg's streamlined, multi-touch sequencer for the iPad; for recording, editing, and mixing; record tracks in CD audio quality; edit music with the key and sample editors; included mixer and audio effects; unlimited audio and MIDI tracks depending on the device used; record up to twenty-four tracks simultaneously; more than seventy virtual instrument sounds based on HALion Sonic; MixConsole with more than ten effect processors; more than three hundred MIDI and audio loops; virtual keyboard and virtual drum pads; sample editor and key editor; export projects to Cubase, Dropbox, SoundCloud, AudioCopy, and e-mail; Core Audio and Core MIDI compatible hardware supported; sequence other Core MIDI apps for MIDI recording only and run app simultaneously via background audio; audio import from iTunes music library, AudioPaste, WiFi server, and iTunes File Sharing; audio mixdown and MIDI export; Apple Core Audio framework; supports AudioCopy and General Pasteboard; Audiobus supported input, output; Apple Core MIDI compliant; Inter-App Audio support; (iOS 5.1 or later).

Dropbox (https://www.Dropbox.com) Lets user bring photos, docs, music files, and videos anywhere and share them easily; access any file saved to Dropbox from all computers, devices, and the web; two GB of space for free; share a link to files; no more attachments; add files to Favorites for fast, offline viewing; (iOS 7.0 or later).

energyXT (http://www.energy-xt.com/index.php?id=0117) Create and record; drag and drop tracks into DAW as MIDI and/or audio files; Apple Core MIDI compliant; (iOS 4.3 or later)

FL Studio Mobile (http://www.image-line.com/flstudiomobile/ipad.php) Create and save multi-track music projects; load projects into the desktop version and take them to the next level; includes 133 high-quality studio-recorded instruments in all musical styles including classic, jazz, rock, electronic, and much more; track editor; step sequencer; piano roll editor; real-time effects; supports AudioCopy and General Pasteboard; Audiobus supported input, output; Apple Core MIDI compliant; Virtual MIDI port; (iOS 5.1 or later; Android 2.3.3 and up).

FourTrack (http://www.sonomawireworks.com/iphone/fourtrack/) Multitrack Recording: use bounce for more than four tracks; sixteen-bit, 44.1-kHz recording quality; Pan Control: move tracks from left to right; Timeline: seek to anywhere in a song instantly; Latency Compensation: accurate to within one ms; Compressor-Limiter: fattens sound of output mix; Bounce: mix song to track one and two of a new song to record many tracks; Metronome: select tempo by number or tapping; record along with real drums, including beats by Jason McGerr; MasterFX: sweeten the sound of recordings with a compressor-limiter and a four-band parametric EQ; File Import: import audio by opening an audio e-mail attachment or by dragging audio files into FileImport area in iTunes; supports WAV, MP3, and AIFF formats; enable/disable input monitoring; jam along with favorite songs while iTunes Music app is playing; GuitarJack Model 2 Control Panel; record in stereo; TaylorEQ; ARM optimized audio engine; higher-resolution graphics; Dropbox Sync; supports AudioCopy and General Pasteboard; (iOS 7.0 or later).

FreEWI (http://www.audeonic.com/#freewi) Designed as a free add-on for MIDIBridge for EWI players; map MIDI events between apps; Apple Core MIDI compliant; MIDIBridge is required; Virtual MIDI port; Universal app; (iOS 4.2 or later).

GarageBand (http://www.apple.com/ios/garageband/) Collection of touch instruments and a full-featured recording studio; use multi-touch gestures to play pianos, organs, guitars, drums, and basses; tap out beats with an acoustic and electronic drum kit; record voice using the built-in microphone and apply sound effects; re-create legendary guitar rigs with nine amps and ten stompbox effects; sampler; record compatible third-party music apps into GarageBand using Inter-App Audio in iOS 7; smart instruments; conduct an entire string orchestra with one finger using Smart Strings; tap chords to create keyboard grooves with the Smart Keyboard; strum chords on an acoustic and electric Smart Guitar, trigger finger picking patterns for popular chords, or switch to Notes view; groove with a variety of Smart Basses using upright, electric, and synth sounds; drag drum instruments onto a grid to create own beats with Smart Drums; start a jam session with friends; play or record live over WiFi or Bluetooth; tempo, key, time signature, and chords automatically sync to the bandleader; jam with any touch instrument or live instruments like electric guitar or voice; bandleader automatically collects everyone's recordings so they can be mixed as a song and shared; arrange and mix song with up to thirty-two tracks using touch instruments, audio recordings, and loops; use the note editor to adjust or fine-tune any touch instrument recording; trim and place musical regions; use the mixer to fine-tune each track's volume; solo or mute any track or adjust pan, reverb, and echo; choose from more than 250 professionally prerecorded loops as a backing band; share songs; keep GarageBand songs up to date across all iOS devices with iCloud; create custom ringtones and alerts for iPad, iPhone, or iPod Touch; share songs directly to Facebook, YouTube, and SoundCloud; e-mail songs right from GarageBand; export song and add it to the iTunes library on Mac or PC; share GarageBand projects directly between iOS devices using Airdrop for iOS; requires one-time in-app purchase for the complete collection of GarageBand instruments and sounds; Audiobus supported output; supports iOS Inter-App Audio; Apple Core MIDI compliant; (iOS 7.0 or later).

Genome MIDI Sequencer (http://www.whitenoiseaudio.com) Pattern-based MIDI sequencer; control MIDI gear and apps; sequence single patterns or a sixteen-track song; MIDI Learn; Virtual MIDI port; Apple Core MIDI compliant; (iOS 4.3 or later).

Hokusai Audio Editor (http://www.wooji-juice.com/products/hokusai/) Multi-track audio editor for iPhone or iPad; record or import a track; full cut, copy, paste, and delete; suite of filters and special effects available; edit many tracks side-by-side, mix them together, and export to WAV or MP4 format; transfer files to computer via USB or Dropbox or send to another app on device; supports AudioCopy and General Pasteboard; (iOS 6.0 or later).

ILRemote (http://www.image-line.com/plugins/Tools/IL+remote/) App is a virtual MIDI controller application for FL Studio (11.1+) or Deckadance 2 (2.3+); (iOS 5.1 or later).

Instrumental (http://www.quantumclockwork.net/instrumental/) Use iPad with MIDI software or hardware as a drum pad, Wicki-Hayden Hex-Keyboard, or full-feedback mixing desk; turns iPad into a flexible MIDI input device for use with any MIDI software running on Mac or Windows PC; with additional hardware such as IK Multimedia's iRig MIDI can also control MIDI hardware; Apple Core MIDI compliant; Virtual MIDI port; (iOS 4.2 or later).

iRig (http://www.ikmultimedia.com/products/cat-view.php?C-mobile) Audio-recording app for iOS device; professional recording tool; Audiobus integration, AudioCopy/Paste, digital audio support, Retina display; intuitive and practical editing functions; export options to fit audio needs; companion to iRig Mic; capture podcasts, interviews, concerts, and any sound out in the field or at home; intuitive editing tools to cut, crop, or loop audio; effects processors to optimize tone for speech or music, clean up background noise, brighten voice for clarity and presence, smooth voice in harsh or overly bright recordings, or change the speed of tracks without affecting their pitch; supports digital audio input devices using any standard thirty pin/Lightning audio interface; organizes recordings by date and time and also tags them with a location; export files via e-mail, WiFi, FTP, SoundCloud, or iTunes File Sharing in a variety of sizes and formats; record with iRig Recorder; one-touch recording with real-time monitoring; recording time is only limited by the storage space on iOS device; eight intelligent effects processors automatically optimize recordings: (1) Optimize Level: automatically adjusts for optimal volume; (2) Optimize Tone: automatically adjusts for optimal equalization; (3) Cleanup: automatically cleans up the background noise in recordings; (4) Brighten Voice: automatically increases clarity of speech and vocals; (5) Smooth Voice: automatically smoothes harsh or overly bright vocals; (6) Speed Up without altering the pitch; (7) Slow Down without altering the pitch; (8) Change Pitch: raises or lowers the pitch of the recording without changing the duration; (iOS 6.1 or later).

iSequence **HD** (http://beepstreet.com/isequenceipad) Music creation studio; intuitive eight-track sequencer; many instruments; flexible mixer with DSP effects; program and record professional loops, beats, and melodies; control every aspect of music project; fluid workflow; record tracks and control movement in real time; edit using step sequencer and automation editor; switch between instruments, tracks, and views without stopping playback; compose, jam, and mix at the same time; Apple Core MIDI compliant; (iOS 4.1 or later).

JACK Audio Connection Kit (http://www.crudebyte.com/jack-ios/) System that connects the music and audio world on iOS devices; currently does not work on iOS 7; (iOS 5.0–6.1.6).

Koushion MIDI Step Sequencer (http://www.koushion.com) Step sequencer that integrates with DAW software and iOS apps that support background MIDI; MIDI Clock Sync; Apple Core MIDI compliant; (iOS 7.0 or later).

Lemur (https://liine.net/en/products/lemur/) Legendary multi-touch MIDI/OSC controller; any software or hardware that receives MIDI or OSC can be controlled by Lemur; control DJ software, live electronic music performance software, studio production software, VJ software, visual synthesis software, stage lighting, and more; Virtual MIDI port; Apple Core MIDI compliant; (iOS 5.1.1 or later).

Line 6 Mobile POD (http://line6.com/sonicport/) App is a companion to Line 6 Sonic Port or Mobile In, which is required for full functionality; Audiobus supported input; (iOS 7.1 or later).

Little MIDI Machine (http://syntheticbits.com/index.html) App is an analog-style MIDI step sequencer for Core MIDI and/or the Line 6 MIDI Mobilizer interface; Apple Core MIDI compliant; MIDI Clock Sync; Virtual MIDI port; (iOS 5.0 or later).

LiveRig MIDI Controller (http://www.pullribbon.com/apps/liverig/) Remotely manipulates DAW, plug-ins, and sounds or effects on computer while playing without having computer in front of them or wires; Apple Core MIDI compliant; Virtual MIDI port; (iOS 5.0 or later).

Logic Remote (http://www.apple.com/logic-pro/) Companion app for Logic Pro X, MainStage 3, and GarageBand on the Mac; designed to take full advantage of multi-touch on iPad; offers new ways to record, mix, and perform instruments from anywhere in the room, turning iPad into a keyboard, drum pad, guitar fretboard, mixing board, or transport control; (iOS 7.0 or later).

Master Record (http://audio-mastering-studio.blogspot.com/2013/07/master-record-for-ipad.html) Recording app with controls and effects like analog tape recorders to give music warmth and natural-sounding, professional-grade quality; can use any external audio source for recording; can be used as effect or output in Audiobus; can upload audio files for final processing or before mixing; cut any part of track; fade-in and fade-out; built-in effects; send track to another audio application thru "Open in" option or use audio clipboard; web access service to share files in WiFi network; Audiobus supported effects, output; (iOS 4.3 or later).

Mastering (http://www.howieweinbergmastering.com) Users go behind the scenes and in the studio with Howie as he shares decades of mastering engineering experience; (iOS 7.0 or later).

Meteor Multitrack Recorder (http://www.4pocketsaudio.com/product.php?p=8) Digital multi-track recorder designed for the iPad; features up to twelve tracks of high-quality stereo, mono, and MIDI audio; built-in mixer and multi-effects processor; Apple Core Audio; Apple Core MIDI compliant; (iOS 6.1 or later).

Microphone + Recording (http://247apps.net/247apps/24_7_Apps.html) Turns iOS device into a microphone; can plug the output of device's headphone jack into powered speakers, a mixing console, home stereo, etc.; (iOS 4.3 or later).

MIDI Bridge (http://www.audeonic.com/#MIDIbridge) Comprehensive MIDI tool; Virtual MIDI patchbay/router/manipulator that interconnects all MIDI interfaces including external, virtual, and network on an iOS device; (iOS 4.3 or later).

MIDI Designer Pro (http://MIDIdesigner.com) Design own MIDI Controller; MIDI rig to control hardware and software synths, effects, and DAWs; pedalboards feature; hardware-software hybrid; Virtual MIDI port; Apple Core MIDI compliant; (iOS 5.0 or later).

MIDI Mandala (Intelligent Gadgets) User selects a musical scale that is mapped to the mandala; Wireless MIDI controller; Virtual MIDI port; Universal app; (iOS 5.1 or later).

MIDI MIND (http://www.dyslexiasoft.com) MIDI interactive note device; allows a MIDI input to produce a multiple output of harmonies and rhythm based on defined parameters; Virtual MIDI port; (iOS 7.0 or later).

MIDI Monitor (http://iosMIDI.com) Test and monitor MIDI hardware; Apple Core MIDI compliant; (iOS 4.2 or later).

MIDI Player (http://www.theMIDIplayer.com) Audio track recording and custom song mixing; tools for the solo performer or backing tracks; Apple Core MIDI compliant; (iOS 4.3 or later).

MIDI Studio Pro (http://www.wiksnet.com) MIDI controller app; Virtual MIDI port; for home or professional studio; Apple Core MIDI compliant; (iOS 4.3 or later).

MIDI Surface 2 (http://www.audiofile-engineering.com) MIDI control surface; audio utilities; technical support; Apple Core MIDI compliant; (iOS 4.3 or later).

MIDI Tool Box (http://mtb.artteknika.com/index_en.html) Integrated app for managing and controlling MIDI devices and data; Apple Core MIDI compliant; (iOS 4.2 or later).

MIDI Touch (http://iosMIDI.com) Create custom controllers by placing knobs, sliders, and other controls onto the screen; Apple Core MIDI compliant; (iOS 5.0 or later).

MIDI Wrench (http://www.crudebyte.com) For daily MIDI setup tasks; connect a MIDI keyboard, MIDI sound expander, or any other MIDI device to iPad, iPhone, or iPod Touch by using the Apple USB camera adapter or any other adapter supported by Apple; visualize MIDI messages sent by the connected MIDI device; send MIDI messages to the connected MIDI device; tool to find problems in MIDI setup or what kind of messages various MIDI devices are sending; supports JACK Audio/MIDI Connection Kit; (iOS 4.3 or later).

MIDI-to (http://www.MIDI-to.com) Wireless DJ MIDI controller designed for Serato Scratch Live; does not need additional software to work; Apple Core MIDI compliant; (iOS 4.2 or later).

MIDIBridge (http://www.audeonic.com/#MIDIbridge) App is a virtual MIDI patchbay/router that interconnects all MIDI interfaces (external, virtual, and network) on an iOS device; Apple Core MIDI compliant; (iOS 4.3 or later).

MIDIflow (http://www.MIDIflow.com) Allows user to send MIDI from app to app; can sync apps with each other or send MIDI parts from a sequencer app to different synth apps; can route the MIDI from a keyboard to apps and assign different key zones to them; all MIDI transfer can be monitored in order to find problems or to learn what MIDI is doing; (iOS 7.1 or later).

MIDImorphosis (http://www.secretbasedesign.com/apps/MIDImorphosis) Captures the pitch of incoming audio and converts it into MIDI notes that can be used to control iOS synthesizers or external equipment; supports both monophonic and polyphonic playing; support for Core MIDI, Virtual MIDI, DSMIDI WiFi, Audiobus, and background audio; connections to laptop and desktop computers can be made using WiFi or with MIDI adapter cables; app is designed to work best with a guitar or bass; (iOS 6.1 or later).

MIDIPads (http://MIDIpads.com) Professional and fully configurable drum pad and MIDI controller for Network MIDI, Virtual MIDI, and hardware MIDI interfaces; can take iPad on stage and trigger sounds wirelessly; Apple Core MIDI compliant; MIDI Learn; (iOS 4.3 or later).

MIDIVision (http://www.audeonic.com/#MIDIvision) App is a real-time MIDI capture, filter, and analysis tool; Virtual MIDI port; Universal app; (iOS 3.0 or later).

MINI-COMPOSER (http://akamatsu.org/aka/ios/apps/mini-composer/) Sixteen-step sequencer; start/stop sequence; multi-touch note input; thirty-two polyphonic tones plus one drum track; four waves (saw, triangular, square, sine); four drumbeat loops; drums on/off; random notes; clear notes; WiFi sync play; (iOS 3.0 or later).

Mixcraft 6 Remote (http://acoustica.com) Control Mixcraft 6 recording software from iPhone, iPad, or iPod Touch; (iOS 5.0 or later).

Mixtikl 5—Generative Music Mixer (http://intermorphic.com/mixtikl/) Generative music and audio loop mixer and cell sequencer; includes twelve tracks and 380 generative templates to play with, edit, and arrange; can add in own audio loops; record mix to audio, MIDI, or tweet/e-mail; Audiobus supported input; supports AudioCopy and General Pasteboard; (iOS 5.1 or later).

Mobile Music Sequencer (http://usa.yamaha.com/products/apps/mobile_sequencer/?cnt=iT) Combine a range of phrase patterns and create musical compositions intuitively, following the flow of composition, from phrases to sections and from sections to songs; sketch the outline of a composition; use the ingredients to craft songs on a Yamaha synthesizer or in Steinberg Cubase; Audiobus supported input; supports iOS Inter-App Audio; (iOS 7.0 or later).

Modular V Controller (http://www.4littlefonzies.nl/#app3) MIDI controller app for the Arturia Modular V virtual instrument; Apple Core MIDI compliant; (iOS 5.0 or later).

MultiTrack DAW (http://www.harmonicdog.com) Digital audio workstation; up to twenty-four audio tracks; record multiple takes, harmonies, solos, and experiments; Audiobus supported output; Apple's Core Audio framework; supports iOS Inter-App Audio; (iOS 5.1.1 or later).

MultitrackStudio for iPad (http://www.multitrackstudio.com) Audio/MIDI multi-track recording app; high-quality audio effects including a guitar amp simulator; both audio and MIDI tracks can be edited; MIDI editing features include piano roll, drum, and score editors; up to sixteen audio/MIDI tracks; audio tracks can be mono or stereo; Editors: audio, piano roll, drum, score, MIDI controllers, timesig/tempo, and song; one effect return section; one master section; three effect slots per mixer section, two for MIDI tracks; Effects: Automated Fader, Chorus, Compressor, DeEsser, Echo, EQ, Flanger, Guitar Amp, Master Limiter, Noise Gate, Phaser, Pseudo Stereo, Reverb, Saturator, Tremolo, Tuner, and Vibrato; MIDI instruments: MultitrackStudio Instruments (General MIDI compatible), SoundFont Player, and one Core MIDI output device; supports Inter-App Audio effects and instruments (iOS 7); supports Audiobus; acts as output in Audiobus app; MIDI sources: onscreen keyboard (keyboard, drum, and various string layouts) and one Core MIDI input device; Audio sample rate: 44.1 or 48 kHz; imports WAV, AIFF, MP3, MIDI, and various other audio file types; exports WAV and MIDI files; import/export audio/MIDI via AudioShare, General Pasteboard, or iTunes File Sharing; imports/exports ZIP file containing all files needed to open song; compatible with MultitrackStudio for Windows/OS X; (iOS 6.1 or later).

Musaico (http://www.musai.co) Streamlined interface for recording and remixing music; record, loop, layer, and remix sounds in real time from layering guitar parts to looping rap beats; supports AudioCopy and General Pasteboard; Universal app; (iOS 4.3 or later).

Music Studio (http://www.xewton.com/musicstudio/overview/) Complete music production environment; combines a piano keyboard, 118 studio-quality instruments with sustain; fully fledged 127-track sequencer; extensive note editing; reverb; real-time effects; user-friendly interface; Apple Core Audio; supports AudioCopy and General Pasteboard; Audiobus supported input, output; Apple Core MIDI compliant; Virtual MIDI port; (iOS 5.1 or later).

Musk MIDI Keyboard (https://sites.google.com/site/muskapps/keys) Multi-touch keyboard; eight octaves; send notes to other MIDI apps or keyboards; receive notes from external MIDI sources; track MIDI notes and channels on keyboard; load sound libraries; change MIDI channels to different sounds; highlight individual channels; iOS 7 voices; (iOS 6.0 or later).

Musk MIDI Player (https://sites.google.com/site/muskapps/player) MIDI player; add sounds and MIDI files; load any MIDI file and sound library; play MIDI files up to eight times slower or faster; follow the MIDI notes on the keyboard and distinguish different MIDI channels; read along the lyrics; app can speak the lyrics during playback (requires iOS 7 voices); app can announce the bar and beat numbers; mute or solo tracks or use them to navigate; supports iOS Inter-App Audio; (iOS 6.0 or later).

n-Track Studio (http://en.ntrack.com/ios-multitrack-studio.php) Audio and MIDI multi-track recorder; turns iPhone, iPad, or iPod Touch into a recording studio; can record and playback an unlimited number of audio and MIDI tracks, mix them during playback, and add effects; edit, cut, copy and paste, zoom, and drag waveforms; Apple Core Audio; Audiobus supported output; supports iOS Inter-App Audio; (iOS 6.0 or later).

NanoStudio (http://www.blipinteractive.co.uk) Recording studio app with virtual analog synths; sample trigger pads; comprehensive sequencer; sample editor; mixer; multiple effects; supports AudioCopy and General Pasteboard; Audiobus supported input, output; Apple Core MIDI compliant; MIDI Learn; (iOS 5.0 or later).

Noisepad (http://www.noise-pad.com) Soundboard, sequencer, and live set DJ app; load own samples or use the in-app shop; create beats; real-time effects; MIDI Clock Sync; MIDI Learn; Virtual MIDI port; Audiobus supported input; Apple Core MIDI compliant; (iOS 5.0 or later).

Peter Vogel Fairlight CMI (http://www.fairlightinstruments.com.au/ios/) Experience music making 1980s style with retro app; modeled on the legendary Fairlight Computer Musical Instrument; preloaded with more than five hundred CMI Series II sounds and more than one hundred Series III sounds; Audiobus supported input, output; (iOS 5.0 or later).

PixiTracker (http://warmplace.ru/soft/pixitracker/) High-quality sixteen-bit sampler; unique sounds; pattern-based sequencer with unlimited number of patterns; sound recorder; MIDI keyboard support; WAV export/import; WiFi export/import; iTunes File Sharing; AudioCopy/Paste; Audiobus supported input; JACK Audio/MIDI Connection Kit supported app; (iOS 4.3 or later; Android 2.3 and up).

Portastudio (http://tascam.com/product/portastudio/) App based on the Porta One recorder that revolutionized recording in 1984; records up to four tracks with a vintage vibe; Apple Core Audio; (iOS 4.2 or later).

PortManager (http://iconnectivity.com) Control iConnectMIDI converged MIDI interface; Apple Core MIDI compliant; (iOS 5.0 or later).

ProRemote (http://www.folabs.com/proremote.html) Thirty-two-channel touch-sensitive control surface; control Audio applications wirelessly using existing WiFi network; (iOS 3.2 or later).

Protein Der Klang (http://proteinderklang.com) Musical sampler designed for public performances; Audiobus supported input; (iOS 5.0 or later).

Reforge—Waveform Editor (http://www.audioforge.ca/reforge.php) Edit a waveform directly by touching it; cut, copy, and paste within files or paste from another file; trim unwanted parts; Apple Core Audio; supports Audio-Copy and General Pasteboard; (iOS 7.0 or later).

Remaster (http://audioforge.ca/more.php) Seven-band equalizer, normalizer; render to file; export setting to use in Equalizer; export settings as PDF; Apple's Core Audio framework; Audiobus supported effects; supports Audio-Copy and General Pasteboard; (iOS 7.0 or later).

Rode Rec (http://www.rodemic.com/software/iOS+Apps) Record, edit, and publish broadcast-quality audio directly from iOS device; (iOS 7.1.or later).

S1MIDITrigger (http://s1products.info) Customizable touch-screen MIDI controller that can replace hardware devices; connect wirelessly to audio workstation applications on PC or Mac or use Line 6's MIDI Mobilizer hardware add-on; Apple Core MIDI compliant; (iOS 5.1 or later).

Sample Lab (http://samplelabapp.com) Full-featured sampler, editor, and sequencer; professional set of features; multi-touch interface; Virtual MIDI port; MIDI Clock Sync; supports AudioCopy and General Pasteboard; (iOS 6.1 or later).

SampleTank (http://www.ikmultimedia.com/products/cat-view.php?C= family-sampletankios) Master-quality MIDI sound and groove workstation for the mobile artist; provides hundreds of world-class instruments and patterns including piano, organ, drums, bass, guitar, strings, synths, percussion, vocals, and many more; (iOS 5.1 or later).

SampleWiz (http://www.wizdommusic.com/products/samplewiz.html) Three sampling modes: classic, granular, and modern; supports AudioCopy and General Pasteboard; Audiobus supported input, output; Apple Core MIDI compliant; MIDI Learn; Virtual MIDI port; (iOS 4.3 or later).

Samplr (http://www.samplr.net) Make music and play with sound in a new and intuitive way by touching the waveform on the screen; explore the sound's melody and texture using the different play modes and create music compositions with the gesture recorder; Audiobus supported input; supports AudioCopy and General Pasteboard; (iOS 5.0 or later).

SimpleMIDIPad (http://www.naughtypanther.com) MIDI CC source for driving software or hardware synths and effects and remote control of a DAW; each of the four pads sends X, Y, and Z MIDI control messages with user-selectable destination addresses; Apple Core MIDI compliant; (iOS 4.2 or later).

Sinusoid (http://www.humbletune.com/sinusoid/) Eight-bit music tracker/sequencer sound toy; four channels, two sine, one square, and one drum channel; sound controls are attack, decay, and tremolo; copy, paste, save, and load; inspired by retro games and old, almost forgotten trackers; Audiobus supported input; (iOS 4.3 or later).

Sir Sampleton (http://www.softofttechech.com) Sampling musical keyboard similar to Casio and Yamaha samplers from the 1980s; allows user to record sounds through the microphone and then play them on dual piano keyboards accompanied by a rhythm machine; Virtual MIDI port; Audiobus supported input, output; Apple Core MIDI compliant; (iOS 5.0 or later).

Sonic Logic (http://www.soniclogicapps.com) Edit and control MIDI controller setups and DAWs from iPad; Apple Core MIDI compliant; Virtual MIDI port; (iOS 5.1 or later).

SpaceSampler (http://www.spacesamplerapp.com) App is for audio convolution; record and share; apply filters from real acoustic spaces and equipment; supports AudioCopy and General Pasteboard; (iOS 6.1 or later).

StepPolyArp (http://laurentcolson.com/steppolyarp.html) Real-time MIDI arpeggiator to control MIDI instruments, sequencers like Logic, Cubase, Live, and all other MIDI sequencers, or other virtual instruments installed on the same iPad; arpeggiator can automatically generate melodic patterns from notes or chords played in real time; Virtual MIDI port; Audiobus supported input; Apple Core MIDI compliant; MIDI Clock Sync; (iOS 5.1.1 or later).

Synergy Studio (http://www.4pocketsaudio.com/product.php?p=12) Sequencer app; by tapping on the grid interface users can lay down a series of notes and patterns of up to sixty-four notes to create a musical sequence; supports AudioCopy and General Pasteboard; Audiobus supported input slot; Apple Core MIDI compliant; (iOS 5.1 or later).

Tabletop (http://retronyms.com) Modular environment; mix and match from more than thirty devices including samplers, mixers, effects, sequencers, and more; supports AudioCopy and General Pasteboard; Apple Core MIDI compliant; supports iOS Inter-App Audio; (iOS 6.1 or later).

TB MIDI Stuff (http://www.thiburce.com/TBStuff/?page_id=664) Full-featured modular MIDI and OSC control surface; design pages with sliders, knobs, faders, drum pads, XY pads, multi-touch XY pads, ribbons, or jog wheels with the built-in editor; three built-in controllers: a keyboard with arpeggiator, a pads controller with configurable pads banks, or a sixteen-channels mixer; Virtual MIDI port; Apple Core MIDI compliant; (iOS 5.1.1 or later).

The Oscillator (http://kymatica.com/Software/Oscillator) Tone generator with two oscillators and one noise generator, each with individual low-pass filters; for testing purposes or to make old-school electronic drone music; all three modules have amplitude control and low-pass filter; oscillators can produce one of four waveforms: sine, triangle, sawtooth, and square; supports Audiobus and JACK Audio connection kit; stream the audio output in real time into other compatible apps; (iOS 5.1 or later).

Thesys (http://sugar-bytes.de/content/products/ThesysIOS/index.php?lang=en) MIDI step sequencer app; Inter-App Audio Support; MIDI file export; integrated synth; Audiobus support; full MIDI Support; MIDI Clock Sync; action section (gatetime, looper, slowdown); pattern sequencer; zoom; (iOS 7.0 or later).

TouchOSC (http://hexler.net/software/touchosc) Modular OSC and MIDI control surface; supports sending and receiving Open Sound Control and MIDI messages over WiFi and Core MIDI inter-app communication and compatible hardware; (iOS 4.3 or later).

TwistedWave (http://twistedwave.com/mobile) Audio editor for iPhone and iPad; Audiobus supported input, output; supports AudioCopy and General Pasteboard; (iOS 5.1.1 or later).

VoKey 2 (http://stonelightpictures.com/VoKEY.html) Sampler; manipulate waveform with touch screen; timeshift to sync notes when playing chords;

record performances and play back; Audiobus, AudioCopy, and AudioPaste support; (iOS 6.0 or later).

Wireless Mixer (http://trajkovski.net) Realistic wireless MIDI mixer controller for computer music application or DAW; emulates a real mixer console with many functions by using MIDI over WiFi; (Android 2.1 and up).

DISCUSSION AND ACTIVITIES:
CREATING MUSIC

$\mathcal{T}$he creative ideas, concepts, and feelings that influence the work of musicians come from a variety of sources. Musical ideas are selected, developed, and generated for various purposes and contexts. The creative choices musicians make are influenced by their expertise and expressive intent. Musicians evaluate and refine their work by trying new ideas, applying appropriate criteria, and persistence. They share creative musical works that convey intent, demonstrate craftsmanship, and exhibit originality. The musician's presentation of a creative work is the culmination of a process of creation and communication. The previous four chapters, chapters 1 through 4, list apps used for creating music, including reading and notating music, composing and arranging music, improvising, and recording music.

TOPICS FOR DISCUSSION

1. As technology grew and society began to communicate on a global level, music also changed. How did these changes affect how music is created and sold?
2. Computers are now being included in the authoring and performing of music. If you made a prediction, in one hundred years, will synthesized instruments have replaced the kind of instruments we play today? Why or why not?
3. How are rhythm, harmony, melody, and variation used in musical works and why?
4. How do musicians generate creative ideas?
5. How do musicians improve the quality of their creative work?
6. How do musicians make creative decisions?
7. How does the composer balance technical skills and planning with spontaneous creativity? Does one revoke or supersede the other?
8. How would creating a film score change your appreciation for those individuals who compose music for movies? Which elements of creating a film score would you find most difficult? Which would you find most rewarding?
9. Identify a song that tells a story. How does the artist or composer use music to tell the story? Do you think music enhanced or detracted from the story? Why?
10. If you were invited by your community to create a piece of music to represent the society, what type of music would you create? What would it express? How would it speak to future generations?
11. In what ways can a piece of music be organized?

12. What are some examples of song form?
13. What examples of rhythm do you find in nature?
14. What is the concept of balance in music?
15. What is the difference between reading music and playing by ear?
16. What is the relationship between form and function in music? Comparatively, how has the relationship between form and function shifted in the contemporary aesthetic?
17. What is the value of musical notation?
18. What other forms of musical notation exist besides Western music notation?
19. When is a creative work ready to share?
20. Why do people create and perform music?

LEARNING ACTIVITIES

1. Compare several music theory, note identification and sight-reading, and scales and chords apps to find which are the most helpful and easiest to use. Practice note and rhythm identification, sight-reading, and identifying music symbols, intervals, key signatures, scales, and chords using the apps.
2. Create a melody using one of the touch instruments in the Garage-Band app and record it. Add a groove to the melody using loops and experiment with the different software instrument sounds.
3. Create a composition project that includes several software instruments, digital audio, and loops using a Digital Audio Workstation (DAW) app.
4. Choose one of the sheet music apps for organizing music scores, songs, and/or lyrics.
5. Download public domain pieces of sheet music from the Internet and practice sight-reading them from a device with a sheet music reader app.
6. Create a file using a computer-based music notation program such as Finale or Sibelius and save it to SongBook (Finale) or Avid Scorch (Sibelius). Transfer files by using iTunes File Sharing, sending an attachment via e-mail, or uploading files from a computer to a cloud service and downloading them to the app.

Part II

PERFORMING MUSIC

• 5 •

Singing and Ear Training Apps

CHILDREN'S SONGS APPS

35 Sing Along Songs (http://pinkfong.com/ssbooks.com_us_android_google-market.html#tab2) Includes thirty-five children's sing-along songs; create playlist by selecting favorite songs; learn the lyrics while singing the songs; can listen to songs even when the screen is locked; doesn't require any network connections; (iOS 4.3 or later; Android 2.2 and up).

ABC Song Sing Along HD PRO (http://www.tapfun.com/abc-song-sing-along.html) Sing along and pop the balloons to learn the song; includes three versions of "The ABC Song"; for each version of the song, balloons float up and user can touch each one in sync with the song; helps kids learn the song and recognize the alphabet; three versions of the song include a guitar version, a young girl singing, and an older woman singing; floating balloons are timed to each version of the song so that kids can learn their letters and the lyrics to the song; designed for babies, toddlers, and children of all ages; engaging and educational; visually and audibly exciting experience; complements and supplements school curricula; (iOS 5.0 or later).

DooDah (http://www.gadjetts.com/DooDah) Drag character around the staff to play song; double tap for spins and giggles; choose favorite color and expression; (iOS 3.2 or later).

English Alphabet Rap (http://www.thrass.co.uk) Catchy rap music for learning the lowercase and uppercase letters of the English alphabet; Music Mode and Speech Mode options; Whole Alphabet Section or Part Alphabet Section; (iOS 4.0 or later).

Kids Karaoke Machine (http://kidsbestplace.com/) Each song is played by a band of animated animals with its own background atmosphere, animations, interactions, music, and adaptive song text; for children of various ages to listen, sing, learn, and enjoy; includes "If You're Happy and You Know It," "Somewhere Over the Rainbow," "Head Shoulders Knees and Toes," "Hush Little Baby," "Itsy Bitsy Spider," "London Bridge Is Falling Down," "Super-califragilisticexpialidocious," "Yankee Doodle," and "Twinkle Twinkle Little Star"; (iOS 4.3 or later).

Kids Song Machine (http://www.kidssongmachine.com) Popular interactive application; for parents to share with their children; songs that inspire kids to listen, sing, and play with music; creative design, fun animations, and upbeat tunes; learn the harmony and rhythm while learning the lyrics; (iOS 6.0 or later).

Learn to Sing Chinese Nursery Rhymes (http://www.bblife.com.cn) Learn Chinese nursery rhymes; improve language fluency, presentation skills, and sound sensation; improve child's emotion and intelligence quotients; familiar and easy to sing; rhymes stimulate kids to sing along with the tune and synchronous words; (iOS 4.3 or later).

Motion Songs—Kids' Songs with Motions (http://motionsongs.com/) Fun and child-friendly app contains ten videos where character Elizabeth performs popular kids' songs with motions; meant to provide entertainment for children as well as give parents a way to get an insight on the pedagogic tools that daycares, kindergartens, and schools use to develop children's motor, verbal, and musical skills; designed with a simplicity that makes it easy to use for children of all ages; subtitles; Loop option; Play All Songs option; (iOS 4.3 or later).

Orff Music for Kids 1—6 (Hongen Education and Technology Co., Ltd.) Effective method of studying with the aid of musical rhythm; designed by Orff music education systems; combines rhythm with reading, percussion tapping, musical games, singing, dancing, small musicals, playing Orff instruments, and more; (iOS 4.3 or later).

Sing Sing Together (http://m.e-learn.co.kr/paid/sub_sstogether1.html) Provides the opportunity to practice listening and reading skills through many songs and rhythmic actions; for first-time preschool and elementary-school-aged English learners; for English learners who need various activities including chants and songs; sing and dance along with 3-D video characters; consists of many famous children's songs that children of all age groups can enjoy and sing; a way to learn English for the first time in a fun and easy manner; (Android 2.1 and up).

Singing Fingers (http://singingfingers.com/) Finger paint with sound; touch the screen while making a sound and colorful paint appears; touch the paint to play back the sound again; paint own musical instrument and experiment with sounds; includes help screens and examples; save and load creations; pitch-tracking algorithm can tell what note singer is singing; the same note in different octaves gives the same color; sing a scale to get a rainbow; (iOS 3.2 or later).

CHOIR APPS

Choir and Organ (www.pocketmags.com) International voice of the choral and organ worlds; essential news and previews; topical features on new and restored instruments; in-depth interviews with choirs, choral conductors, and composers; for organists, choral directors, singers, organ builders, listeners, or those working in the publishing or the recording industry; insights into the lives and views of leading organists, choral directors, and composers; illustrated features on newly built and restored organs; features a new young composer commissioned work in every issue, with the score available for download and performance; reviews of the latest CDs, DVDs, and sheet music; listings of recitals, festivals, and courses; free monthly e-bulletins with breaking news, editor's concert picks, and competitions; (iOS 4.3 or later).

Harmony Voice (http://www.virsyn.net/mobileapp/index.php?option=com_content&view=article&id=8&Itemid=8) Transform voice into a choir with up to four voices; pitch shifter and harmonizer with professional features including automatic tuning correction; real-time visual intonation display gives feedback of the tunes sung; will add up to four voices according to the chords played with the piano keyboard; can play chords automatically to enrich the sound of user's voice; voice character can be adjusted; manual harmonization by playing chords with up to four notes while singing; automatic harmonization by playing bass note and app adds appropriate chords; directly play the tunes for up to four voices with the keyboard; select key note and scale for harmonization and pitch correction; harmonies can be in just or tempered tuning; mix voice with harmonized parts; play background track from iTunes library; reverb effect for room simulation; chorus and delay effect; Audiobus support (input/filter/output); Core MIDI compatible; audio recorder with metronome; Audio pasteboard for exporting to other apps; headphones required to avoid feedback; (iOS 5.0 or later).

Sing Harmonies (http://www.singharmonies.com/) Fun harmonizing tool with which singers can learn to sing vocal harmony by playing and muting individual voices while playing four-part vocal arrangements of "Lean On Me," "Teach Your Children," and "Proud Mary"; (iOS 3.0 or later).

Virtual Chant Instructor (http://agesinitiatives.com/) Developer envisions all Orthodox churches having the texts, music, and trained singers needed to maintain the fullest possible liturgical life; multilingual database, rubrics engine, and applications for church, home, and classroom, as well as translation tools for use in the mission field; software platform will store, organize, and deliver Orthodox Christian liturgical texts and music to support the education of church singers and to facilitate the smooth performance of church services, using current and emerging technologies; (iOS 5.1 or later).

Vocal Warm-Ups for Singers or Choir (http://www.choralresource.com/) Vocal warm-ups designed for solo singers or choirs; warm-ups have been used with elementary school through to college-level singers and choirs; warm-ups are organized into five sets; each set includes a physical warm-up, breath awareness, then five minutes of a variety of mid-range warm-ups; includes a vocal example with a male and a female vocalist demonstrating each exercise; click track gives the tempo; a vocal model accompanies the vocalist(s) for the first exercises, then student or group continues until the end of the exercise; can pause the audio, stop it, or rewind and repeat the exercise; choose a set to work on, or scroll to specific exercises; (iOS 4.3 or later).

EAR TRAINING AND SIGHT SINGING APPS

Absolute Pitch! (Lei Chi Tak) Identify pitch and simple scales; Play Again button; improves sense of pitch; trains listening skills; perfect pitch training; relative pitch training; absolute pitch training; randomly plays a particular sound, then user chooses the pitch of the sound; records the number of correct answers; users find their own level; "Hint" prompt function for beginners; C major and mixed mode; (iOS 3.2 or later).

Better Ears (http://www.better-ears.com/en/better-ears-ipad.html) Formerly known as Karajan—Music and Ear Trainer; educational music and ear training app; helps user grow their musical skills and enhance hearing capabilities; for beginners and music-masters; decide own training routine, when to start, and how often to practice; includes ten different exercises: Interval Recognition, Scale Recognition, Chord Recognition, Chord Progressions, Pitch Recognition, Tempo Recognition, Key Signature Recognition, Interval Music Reading, Scale Music Reading, and Chord Music Reading; choose training sound from six different instruments including acoustic grand piano, bright acoustic piano, electric grand piano, drawbar organ, nylon acoustic guitar, and steel acoustic guitar; comes with a virtual keyboard and virtual guitar fretboard; four preset levels include beginner, easy, normal, and professional; unlimited custom levels; keep Mac and iOS devices in sync; supports routing sound through an external MIDI device to use existing MIDI master keyboard or sound module; includes statistics; shows the relevant Wikipedia article for the current exercise; (iOS 5.0 or later).

Blob Chorus (http://www.lumpty.com/music/music.html) Introduce children to pitch and improve their musical ear; each of the blobs in the chorus sings a note, then King Blob sings his note; aim of the game is to identify the blob that sang the same note as King Blob; will improve child's ability to pick out a note from a sequence; (iOS 4.3 or later; Android 2.3.3 and up).

Do Re Mi Ear Training (http://www.axe-monkey.com) Solfège practice partner; helps develop pitch recognition, memory, and transcription ability by playing musical phrases for the student to repeat back; options for beginners and experts; (iOS 4.3 or later).

Ear Trainer (http://www.dev4phone.com/eartrainer.html) Educational tool designed for musicians, music students, and anyone interested in improving their musical ear; includes more than 230 individual exercises covering intervals, chords, scales, relative pitch, and melody; playable keyboard with studio sound quality and a note view; for all users from beginner to advanced; covers the following categories: Interval Comparison, Interval Identification, Chord Identification, Chord Inversions, Chord Progressions, Scales, Note Relative to Chord, Relative Pitch, and Melody Exercises; each exercise has a short description of its purpose and optional preview; virtual piano keyboard and note view in each exercise helps user visually understand and analyze the notes that have been played; compare the different options after answering; statistics are collected as user progresses through the exercises to follow progress; questions in the exercises are randomly generated and can be done over and over again; (iOS 5.0 or later).

Ear Trainer for Children (http://www.iphone-ipad-iapp.com/ipad.html) Students recognize musical intervals and complete a score; Retina display; Universal app; (iOS 7.0 or later).

Ear Training (http://tonalapps.com/eartraining/) Improve relative pitch; learn major and minor modes, intervals, triads, inversions, and seventh chords; practice recognizing chords and scales by ear with notation; test ability to recognize chords and scales by ear alone; test results are broken down by category to focus on the categories that need improvement; designed to be used with earphones or headphones; (iOS 7.0 or later).

Ear Worthy (http://ixorastudios.com/portfolio-item/ear-worthy/) Comprehensive ear training app; includes an extensive variety of musical listening and identification skills from notes and intervals to scales and chords; includes high-quality piano samples and an intuitive user interface; has four main features that play musical elements; user is given the opportunity to identify what is being played: (1) To identify notes—random notes are played individually; (2) To identify intervals—two notes are played sequentially; (3) To identify scales—a set of notes are played sequentially, ascending, and descending; (4) To identify chords—three to five notes are played simultaneously; (iOS 4.3 or later; Android, varies with device).

EarMan (http://ipad.rogame.com/pages/EarMan.html) App helps improve skills with a structured ear training curriculum and fully customizable practice sessions; (iOS 6.0 or later).

goodEar Chords—Ear Training (http://www.calypso-score.com/goodear .html) For beginners to professional musicians; includes all possible triads and inversions, all possible four-note chords, and twenty different extended/ altered chords; create own exercises; play the exercises on piano or give the answer with multiple-choice; watch progress in statistics; (iOS 6.1 or later).

goodEar Intervals—Ear Training (http://www.calypso-score.com/goodear .html) Ear training for beginners to professional musicians; choose from all intervals between two octaves and play them ascending, descending, or simultaneously; activate "backing chord" to hear the interval in harmonic context; create own exercises; play the exercises on piano or give the answer with multiple-choice; watch progress in statistics; (iOS 6.1 or later).

goodEar Scales—Ear Training (http://www.calypso-score.com/goodear.html) Ear training for beginners to professional musicians; thirty-eight scales including ionian scales, harmonic-minor scales, melodic-minor scales, five pentatonic scales, five hexatonic scales, and seven octatonic scales; play scales ascending or descending; create own exercises; play the exercises on piano or give the answer with multiple-choice; watch progress in statistics; (iOS 6.1 or later).

Music Cubes (http://www.foriero.com/en/pages/music-cubes-page.php) Listen to the cubes and repeat the sequence of notes back; exercise music memory, train, play, and compete with mates; develop tone hearing; develop absolute pitch; concentrate; (iOS 5.1 or later).

Perfect Ear Pro (http://perfectear.educkapps.com/) Easy and fun ear training tool that will help both professional musicians and those who consider music a hobby improve their musical ear; seven exercise types; custom exercises; scale viewer; (Android 2.1 and up).

Pocket Ear Trainer (Polemics Applications LLC) Contains randomized ear trainers for chords, intervals, and scales; tap Play and app will play a random interval, scale, or chord and user selects what type it is; also includes a reference and practice section; read information about the sounds being played and selectively hear certain intervals, chords, and scales; created by a music educator with college music students, band directors, and musicians of all types in mind; learning to recognize intervals by ear is an important part of playing music by ear and an important part of overall musicianship; (iOS 5.1 or later).

Prima Vista Sight Singing (http://android.rene-grothmann.de/primavista.html) Helps students learn sight-singing and how to write down music with melodic dictation; options include the voice (soprano, alto, tenor, bass) and its compass (narrow, normal, wide), the key or a random key for a given number of accidentals, the instrument (violin, French horn), the playing speed, rhythm and melody, rhythm only, or melody only, length of the melody, width of the tune; documentation in English and German; (Android 2.2 and up).

Rhythm Sight Reading (https://sites.google.com/site/sightreadrhythm/home) Trainer with exercises in 2/2, 2/4, 3/4, 4/4, 5/4, 6/4, 7/4, 6/8, 7/8, 9/8 time; exercises can be played at different speeds; rhythms are played with slight sub-accents to emphasize note groups; practice mode; test mode; visual/audio feedback of the rhythmic performance; tap accuracy is measured in seconds; best time is stored for each exercise; more than two hundred fixed exercises; more difficult ones include rests, ties, syncopations, duplets, triplets, quadruplets, and swing eighths; thousands of random exercises; forty difficulty levels; option to add swing to every quarter note meter exercise; metronome with speed of 20 to 240 beats per minute with optional upbeat sound; (iOS 5.1 or later).

Sight Reading HD (http://www.superkiddostudio.com/SightReading.php) Sight-reading app that can train in all key signatures; user-friendly design; learn and practice both treble clef and bass clef notes on eighty-eight keys;

customize practice to any skill level via changing key signatures, accidentals frequency, keyboard range, note selections; enable/disable sound effects and hint option to train ear or let app guide; informative graph generated from history of practice statistics records progress; Retina display; (iOS 4.2 or later).

Sight Reading Trainer (https://sites.google.com/site/sightreadrhythm/home) Improve sight-reading with immediate feedback of timing accuracy; graduated levels from simple to professional; practice anytime and anywhere; advanced features include common and rarer rhythms from two to nine beats per bar including 2/2, 4/2, 2/4, 3/4, 4/4, 5/4, 6/4, 7/4, 6/8, 7/8, and 9/8; plays the rhythm with metronome sound and visible beats; dynamics emphasize downbeats and note groups; tap to follow the rhythm with accuracy displayed immediately; adjust metronome tempo by sliding or keypad; Practice Mode and Test Mode; more than two hundred fixed exercises; more difficult exercises include rests, ties, syncopations, duplets, triplets, quadruplets, and swing eighths; thousands of random exercises; twenty levels of difficulty in 4/4 and twenty levels distributed among the other meters; option to add swing to every quarter note meter exercise; different degrees of swing include light, medium, and hard; classical metronome with speeds up to 240 beats per minute and optional upbeat sounds; sounds of the player, metronome, and taps can be chosen from several percussive instruments; automatic count-off; optional beat numbers in the staff and beat-counting voice function clarify the relationship between the rhythm and the beat; use the microphone as an alternative to the tap button; play the rhythm with hand claps, finger snaps, or an instrument; optional stereo sound; hear taps in the right ear and the metronome and player sounds in the left ear; progress reports for exercises; landscape support to display longer exercises; Universal app; Expert Mode has a higher default tempo and more accuracy is required in tapping; ear training function to help in rhythmic dictation; listen to the rhythm once or several times and write it down or tap it out to compare with the written rhythm; endurance and speeding-up tests; hold tied notes and play legato; (iOS 5.1 or later).

Singsolatido (http://www.absolatido.com/singsolatido/) Sight-singing practice; for anyone who sings in a choir or who has singing lessons; records voice; pointer indicates if voice was in pitch; generates exercises at ten different levels; (Android 2.1 and up).

KARAOKE APPS

Android Karaoke—Sing-Along (http://www.maxdroid.com/) More than one thousand popular karaoke videos; new videos added daily; practice singing anywhere; send song suggestions from the app; easy to use, nothing to download; search by song title or artist; create favorite videos list; karaoke videos of Britney Spears, Beyoncé, Justin Bieber, Adele, Paramore, Christina Perri, Katy Perry, Linkin Park, Brad Paisley, Lady Gaga, Maroon 5, Garth Brooks, the Cure, Depeche Mode, Kenny Chesney, Carrie Underwood, Lionel Richie, Frank Sinatra, Sugarland, Taylor Swift, Pink, Rihanna, Jim Croce, Keith Urban, Kenny Rogers, and more; (Android 1.1 and up).

Disney Spotlight Karaoke (http://spotlight.firstact.com/contact/) Sing and practice favorite songs, put on a show, or make a video; streaming lyrics cue user when to sing; built-in pitch correction; turn song vocals up or down; add voice effects including reverb, tone, and echo; in-app vocal lessons with vocal warm-ups and tips; record performance, then share with friends; access iTunes library; (iOS 4.3 or later).

Hindi Karaoke Sing Along (Sparky) Access Bollywood video karaoke songs on YouTube; use Refresh option in menu to access new songs; search option; (Android 2.2 and up).

Karaoke Anywhere (http://sing.karaokeanywhere.net) Karaoke application with a streaming library of more than ten thousand songs; put together a set list and entertain; output to television with a standard audio/video cable; key changing functionality to fit vocal range; record, mix, and share performances in iTunes compatible format for later playback; share to Twitter and Facebook directly from the application; full headphone monitoring during audio playback with optional reverb and echo vocal FX; purchase subscription or individual songs from library of more than forty thousand tracks in the Digital Download Store; community of users; bulk import or export songs and performances via iTunes File Sharing for backup and playback; compatible with the iRig Microphone; uses the MP3+G ZIP format to display song lyrics and graphics in sync with MP3 quality music; (iOS 5.0 or later; Android 2.3.3 and up).

PocketSing (http://pocketlabworks.com/pocketSing.html) Karaoke singing app with vocal effects; lyrics; sing along with access to iTunes; (iOS 3.2 or later).

Red Karaoke Sing & Record (http://www.redkaraoke.com/) Sing and record; includes the Song of the Day with a different hit every day; thousands of karaoke songs and new weekly productions with the latest from the charts and classic essentials; sing offline without Internet access; record a video while singing and share; sound controls and effects for making recordings; remove unwanted songs; turns device into a portable karaoke machine; (Android 2.2 and up).

Sing Me Something (http://www.singmesomething.com/) Karaoke entertainment; challenge friends' music knowledge by picking a tune, singing it, and asking them to guess the name of the song; create a game with friends or by random; record self singing with help from a sample and lyrics; send song to opponent; wait for them to guess the song; support for English, French, and Spanish languages; (iOS 5.1 or later; Android 2.3 and up).

Sing! Karaoke (http://www.smule.com/apps#sing) Sing favorite songs from a large catalog of top hits; perfect a song and then share it with the world; duet mode; includes songs from many genres, including pop to rock, hip-hop to musicals, country to classics; new and free songs are added frequently; buy a subscription for unlimited access; songbook includes hundreds of new top hits and many more; suggest songs; voice enhancement technology; Top Performances list with singers from all over the globe; (iOS 6.0 or later; Android 4.0 and up).

Soulo Karaoke (http://www.soulo.com) Pro Pitch enhancement helps singer sing in key; video FX and recording to make own music videos; edit and share recordings via YouTube, Facebook, and Twitter; sing along with iTunes library; in-app store; adjustable guide vocal; built-in voice effects include reverb, echo, and tone; Soulo microphone; large selection of new and classic hits; songs added each week; (iOS 4.3 or later).

Spanish Karaoke—Sing-Along (http://www.maxdroid.com/) Sing favorite Spanish songs; more than three hundred of the most popular Spanish and Latin karaoke videos; new songs added daily; practice singing anywhere; send song suggestions; (Android 1.1 and up).

StarMaker: Karaoke + Auto-Tune (www.starmakerstudios.com) Karaoke game with real Auto-Tune; sing, record, and compete with hundreds of hit songs made famous by Justin Bieber, Lady Gaga, Maroon 5, and others; share recordings using Facebook, Twitter, and e-mail; subscribe for unlimited access to sing all the songs; hit songs released every week; (iOS 6.0 or later).

The Karaoke Channel (http://www.thekaraokechannel.com/mobile/) Karaoke app; in-app subscription to more than eight thousand songs; song library includes songs in the style of top-charting artists and music legends; browse top songs in the charts or search by title, artist, or lyrics; high-quality karaoke videos streamed over mobile network or WiFi; listen and sing along to songs with or without lead vocal; access recently played songs and favorites; share songs with friends; advertisement free; listen to thirty-second samples of full karaoke catalog available by subscription; songs added regularly; support for AirPlay with Apple TV; Video Out feature works with supported devices and AV adaptors; genres include pop, rock, RnB/hip-hop, country, Latin, and more; recording of songs is not supported on mobile app; recording feature is available on the Karaoke Channel online; (iOS 4.3 or later; Android 2.2 and up).

VoiceMatch (http://www.WhatIsVoiceMatch.com/) Mobile voice tuner and karaoke app for singers of all levels; know vocal range; like a mobile voice coach; find songs that match singer's own voice; for professional or amateur singers wanting to tune or expand their vocal range; eliminates the need for a pitch pipe; find songs in all music genres; always have a list of songs handy; for singers, actors, musicians, and choirs; (Android 2.1 and up).

OPERA AND CLASSICAL SINGING APPS

100 Greatest Singers (Hon Fai Yiu) Introduces 101 classical singers, including Caruso, Callas, Pavarotti, and more; search with A–Z indexing or in voice typed categories soprano, mezzo-soprano, contralto, tenor, baritone, and bass; each singer has a photo, full biography, and an MP3 song already in the app; songs list table to search the one hundred songs; each singer has two YouTube video links to watch the singer; includes a file to introduce the voice type and how one can determine the voice type; includes art songs and operatic arias; (iOS 6.0 or later).

100+ Greatest Operas (Hon Fai Yiu) For opera lovers, music students studying opera, or beginners trying to understand opera; introduces 110 famous operas by forty-eight composers; each opera has a stage photo, year of composition, and premiere details; includes a detailed description of each opera from Wikipedia, a full opera video, and two excerpt videos from YouTube; synopsis of each opera telling the story; description of each composer; includes operas from the Baroque Era to the twentieth century; opera composers include Verdi, Wagner, Puccini, Donizetti, and many more; operas includes *La Traviata, Der Ring des Nibelungen, Turandot, Lucia di Lammermoor,* and many more; regular length of each is about two hours; 110 operas gives 220 hours of entertainment; opera videos are sung by world-famous singers such as Pavarotti, Domingo, Callas, Sutherland, and more; staged in world-famous opera houses such as Royal Opera House, Vienna State Opera House, Metropolitan Opera House, etc.; (iOS 6.0 or later).

German Diction for Singers (Jeremy Hunt) Interactive app for singers and conductors performing in German; includes more than one hundred audio examples of German vowel and consonant sounds, an IPA guide, and rules for singing in German; built-in voice recorder; use the recorder to check pronunciation of each audio example; (iOS 6.1 or later).

It. Diction—Italian Diction for Singers (Jeremy Hunt) Interactive app for singers and conductors performing in Italian; includes more than one hundred audio examples of Italian vowel and consonant sounds, an IPA guide, and rules for singing in Italian; built-in voice recorder to check pronunciation of each audio example; (iOS 6.1 or later).

Italian Lyric Diction—E & O (https://www.facebook.com/italianlyric diction) Reference for singers and voice teachers; identifies the correct pronunciation of stressed E and O vowels in Italian lyric diction; (iOS 6.1 or later).

Met Opera on Demand (http://www.metoperafamily.org/ondemand/index .aspx) Instant and unlimited access to hundreds of Met performances; subscription streaming service; includes the Met's award-winning *Live in HD* series of movie-theater transmissions; classic Met telecasts; more than three hundred radio broadcasts dating back to 1936; (iOS 4.3 or later).

Opera Classic Music Collection (http://strongwang2012.blog.163.com/) Music player with knowledge; includes portraits of forty-seven great composers; read about them while listening to ninety-nine of the best classical opera compositions; shuffle or order play; repeat mode; music listed by composer, favorites, or all; search by composer or music; (iOS 4.3 or later).

Schott Pluscore Sing-Along (http://www.schott-music.com/pluscore/sing-along) Unique app with adaptable accompaniments; ideal trainer for rehearsing opera arias and other repertoire; offers sheet music that can be set to the individual needs of the singer; add page breaks, select a zoom level, or transpose the music into another key; integrated MIDI Studio and "Virtual Conductor"; can adjust the accompaniment to match interpretation, set the tempo, and change the volume; "Virtual Conductor" allows students to conduct their own accompaniment and then use it for practice; listen to voice part separately or record own interpretation; app offers opera arias for soprano, mezzo-soprano, tenor, baritone, and bass; includes arias by composers such as Verdi, Mozart, Bizet, and Wagner; new sheet music added on a regular basis; (iOS 4.2 or later).

Vocal Tech—Guide to Classical Singing (Jeremy Hunt) Interactive guide to classical singing; intended for students of classical vocal music including opera, oratorio, song literature, etc.; for any student of singing hoping to build a solid technical foundation; discusses phonation, resonance, posture, and breathing; includes vocal exercises with recorded examples and a voice recorder for the student; created by a tenured university professor with a doctorate in voice performance and vocal pedagogy; (iOS 6.1 or later).

POPULAR SONGS AND ARTISTS APPS

Accompanist (http://www.zilibowitzproductions.com/Accompanist/Welcome.html) App will work out the chords to almost any tune user sings; (iOS 6.0 or later).

American Idol (http://www.americanidol.com/app) Companion experience to the singing competition; watch and share on-demand videos, join the conversation, and get the latest Idol news; during live shows can vote for your favorite contestants in real time, view photos, keep up with the judges, and more; featuring host Ryan Seacrest and the current judging panel; (iOS 7.0 or later; Android 2.3 and up).

AutoRap by Smule (http://www.smule.com/apps) Turns speech into rap and corrects bad rapping; maps the syllables of speech to any beat, creating a unique rap every time; tap to rap; Talk Mode: talk into the app and app morphs speech into a legit rap; create original rap songs with freestyle beats or use premium songs from famous rap artists; Rap Mode: rap along to favorite songs, following the lyrics that scroll across the screen; Rappify setting will correct user's flow, snapping syllables to the rhythmic grid of the underlying beat; Rap Battle: challenge anyone to a rap off and weave both component's verses into a single track; headphones recommended; share creations with friends via e-mail, Facebook, or Twitter; new beats and songs added every week; listen to the top rappers and support them with props; earn additional plays; (iOS 6.0 or later; Android 4.0.3 and up).

Celebrity Singers Tile Quiz (http://www.pendrush.com/) Challenges players to guess celebrity singers; includes Bruno Mars, Adele, Katy Perry, Eminem, Jennifer Lopez, Lady Gaga, Madonna, Michael Jackson, Nelly Furtado, Rihanna, Shakira, Tarja Turunen, and more; the more questions answered correctly, the higher the score; two hundred celebrity singers questions, with four possible answers each; top one hundred list for best players; frequent updates; (Android 1.6 and up).

Classic Rock Lyrics Trivia (Sandman Marketing) Lyrics from classic rock artists; guess the correct song; questions are randomized so every game is different; (iOS 6.0 or later).

Logo Quiz—Pop Singer (Logo Quiz) Identify pop singers; more than 480 singers, divided into eight levels; useful clues; each flag has five tips; detailed statistics; best player evaluation; puzzles of varying difficulty; frequent application updates; (Android 2.2 and up).

London A Cappella (Easy Ear Training) Official app of the London A Cappella Festival 2013; listen and explore the music of festival performers including Swingle Singers, the King's Singers, the Magnets Postyr, Rajaton, Retrocity, and the Choir of Clare College, Cambridge; read in-depth articles about the music; original video introductions from the artists themselves; learn about the interactive workshops from world-leading a cappella experts Tine Fris, Christopher Diaz, Ben Parry, Bill Hare, and Erik Bosio; connect with other a cappella fans through the app's social features; (iOS 4.3 or later).

Song Quiz, Guess Radio Music Game (Celso Zlochevsky) Includes thousands of songs; listen to song clips and guess who is playing; genres include rock, pop, and country; (iOS 4.3 or later).

Songify (http://www.smule.com/apps#songify) Turns speech into music automatically; speak into device, and app will turn speech into a song; (iOS 4.3 or later; Android 2.2 and up).

The Great Singers Quiz (desarrollo.app.android) Challenge for music fans and artists; more than 750 singers in twenty-six levels; gives a photo and a name for guessing the singer; score level varies; visual memory game with only one photo and no more clues; (Android 1.5 and up).

The Voice: On Stage (www.starmakerstudios.com) Official game for *The Voice*; sing through favorite songs from *The Voice* to unlock more songs and game levels; see notes and listen with real Auto-Tune; sing, record, and compete with more than two hundred hit songs; share recordings using Facebook, Twitter, and e-mail; start with free tokens and sing to earn more; subscription with new hit songs released weekly, including tracks sung on *The Voice* programs around the world; (iOS 6.0 or later; Android 4.1 and up).

Who Sings It? #1 Hits (http://www.AZTrivia.com/) Music trivia game quizzes knowledge of the artists and groups who have had the biggest hits from 1960 to the present; user is presented with the song title and the year the song was number one; challenge is to choose the artist or group who performed the song; featured artists include Lady Gaga, Elvis Presley, the Beatles, Beyoncé, Jay-Z, Britney Spears, Taylor Swift, the *Glee* cast, Katy Perry, Rihanna, Maroon 5, and many more; other apps in series include songs by decade; (Android 2.2 and up).

SINGING LESSONS AND VOCAL TRAINING APPS

Adictum Singing Lessons (http://www.suonnica.com/asl/) Professionally developed warm-up exercises designed to train any vocal range; exercises use various types of arpeggios in all keys to help boost the student's melodic singing and tuning potential; includes harmony training; helps the singer to develop ear training by recognizing the notes within a predetermined chord; major and minor chord diagrams are included; breathing techniques; how to stay in tune; vibrato; for amateurs and professionals; (iOS 4.3 or later; Android 2.2 and up).

Do Re Mi Voice Training (http://www.axe-monkey.com) Tool for improving singing voice; voice analysis allows singers to see what pitch they're hitting and work on steadiness or vibrato; strengthen passaggio; compare to reference pitches from included keyboard; (iOS 5.0 or later).

Erol Singer's Studio—Voice Lessons (http://www.erolstudios.com/singers-studio-voice-and-ear-training/) Complete voice and ear training program for singers of all levels from absolute beginner to advanced; during the exercises, notes to be sung are shown as blue/green bubbles; blue line shows actual pitch; work on smooth transitions between registers; identifies problem areas; thirty-six lessons including warm-ups, breathing exercises, and exercises to improve tone quality; practice at a slower or faster tempo in any key; male and female examples; exercises go through keys within the student's exact range; app can detect singer's range; (iOS 5.0 or later).

Jeannie Deva Singer World (http://www.jeanniedeva.com) Free multi-media access to vocal tips and warm-ups and singing exercises; for beginners or professional singers; valuable information to help singers achieve their own unique sound and unforgettable vocal performances; (iOS 3.0 or later; Android 2.2 and up).

Learn to Sing—Beginners Tips and Tricks (GR8 Media) Contains 220 easy-to-follow video tutorial lessons on how to sing better; breathing; voice warm-ups; pop vocal exercises; pop singing lessons; vocal training; (iOS 5.1 or later; Android 2.3.3 and up).

Singing Lessons (BPMcorp) Improve vocal skills with more than twenty-five videos showing how to improve singing techniques; (Android 2.2 and up).

Singing Lessons Voice Training (Andys Apps) For anyone looking to improve their vocal skills; watch singing lessons for both beginners and advanced singers; get tips for improving vocal performance; singing techniques are taught by a professional voice coach; several modules such as proper singing methods, care for the voice, proper breathing, and stage performance; learn how to sing in front of an audience; for any age; (Android 2.2 and up).

Singing Secrets (SeedBright) Singing lessons to become a better singer; vocal technique; singing exercises; lessons for kids; warm-ups; how to sing high notes; (Android 2.2 and up).

Singing Teacher (Lubomir Havran BINARTS) Real-time features include singing note analyzer; oscilloscope; voice chart; level meter in dB; stop/start button; noise threshold control; mic boost; (iOS 5.1 or later; Android 2.1 and up).

Singing Teachers on Demand (http://danahartsman.com/) For all singers; scours thousands of programs online and finds the best training videos created by singing professionals; tips and exercises to improve singing skills; chat with other aspiring singers through Twitter to find out which lessons and vocal exercises are working best for them; (Android 2.2 and up).

Singing with the Stars "Voice Lessons" (NextUpApps LLC) Designed to assist all singers and those with singing aspirations to understand the basics; videos to help improve singing; record and listen; vocal exercises; singing tips; how to hit high notes; how to read music; learn the voice types; poll question; notes page; calendar; three mini games; (iOS 5.0 or later).

Virtual Voice Lessons (http://www.virtualvoicelessons.net/home/) Learn how to sing with video singing lessons and vocal training MP3s from Steve Childs; twenty-five complete instructional video lessons from one of the world's top vocal instructors; each lesson is a video from twelve to thirty minutes long, and most lessons include audio exercises as MP3 files; includes six free lessons and more can be purchased; lesson titles include: Introduction to Contemporary Singing, Breathing Part 1: Inhalation, Breathing Part 2: Exhalation, Breathing Part 3: Intercostal Breathing, Understanding Vocal Tone, Understanding Vocal Resonance, Developing Your Vocal Tone, How to Project Your Voice, Controlling Your Voice, Vocal Control for Advanced Breathing, Increasing Your Range Part 1: The Foundation, Increasing Your Range Part 2: Falsetto, Increasing Your Range Part 3: Belt Voice, Vocal Expression, Vibrato, Consonant Connections, Introduction to English Vowels, English Vowels Part 2, English Vowels Part 3, How to Control "EE," Am I Tone Deaf? Singing Dangerously, Singing Dangerously Again, In Sickness and In Health Part 1, and In Sickness and In Health Part 2; (iOS 5.0 or later).

Vocal Mastery (http://www.vocalmastery.com/about.php) Stage 1 is about the function of the voice and includes vocal exercises, warming up, vocal health, and assessing the voice; Stage 2 is about application, where the students apply what they have learned in the exercises into a song, also using a song as an exercise; Stage 3 is about performance, covering many aspects of performance such as how to select material, how to prepare for performances, recordings, auditions, and more; (iOS 4.3 or later; Android 2.2 and up).

Vocal Training (http://www.thunderhillapps.com/support/) Two hundred video lessons to help train the voice; training; practice; correct techniques; how to protect the voice; (iOS 6.0 or later).

VocalizeU (http://vocalizeu.com) Practical app and software program for singers; world-class vocal and musical education; each warm-up, lesson, and evaluation is personalized; caters to singers' individual needs, allowing them to work on areas of their voices at their own pace; patented vocal evaluation technology helps singers determine areas of their voices that need growth and then directs them to customized vocal exercises and workouts; progress can be tracked over time; guides each individual to appropriate workout modules; singers have the option to schedule an online vocal evaluation or complete voice lessons with a live VIP instructor directly from the software; also provides vocal studio tools, educational textbooks, and further music training; (iOS 4.3 or later).

Voice Lessons to the World (http://theapp.voicelessonstotheworld.com/) Episodes created by New York Vocal Coaching's Justin Stoney; mission is to make voice lessons available to anyone, anywhere in the world; articles; practical advice to help singers hone and flex their musical skills; (iOS 6.1 or later).

VoiceSimon (GNN CORP) Voice training game for people who want to brush up their singing skills by themselves; (iOS 5.1 or later).

VoixTek Voice Training (http://www.ronandersonvocals.com/voixtek-app/) Ten exercises; features Ron Anderson's forty years of experience and approach to delivering a better vocal performance; set morning and afternoon practice sessions with reminder alarm; update friends and family on progress; all levels; (iOS 3.2 or later; Android 2.2 and up).

Vox Tools: Learn to Sing (http://vox-technologies.com/vox-tools) Programmed by vocal technique teachers of Vox Technologies; fully guided training section; audio examples for men and women; choose voice type for more accurate and personalized training; personal voice recorder and virtual piano; access to blog with tips; (iOS 5.1 or later).

VOCAL WARM-UPS AND SINGING EXERCISES APPS

ProVocalz (Roarcus LTD) Student creates own library of singing exercises; used in recording studios to prepare singers for sessions; used as an aid for singing teachers during lessons and for their students to practice between lessons; (iOS 4.3 or later).

Singer's Mate (http://www.singersmate.com/) Practice singing anywhere, anytime; practice from lead sheets in MusicXML format; automatically adjusts to singer's vocal range with Transposition Tool; adjust practice parameters for singer's performance; stay on key with visual feedback from the Pitch Detector; (iOS 5.1 or later).

Singing Vocal Warm Ups—Singer's Friend (http://www.singersfriendapp.com/) Warm up and practice on the go with fifteen different scale patterns; choose from bass, baritone, tenor, alto, mezzo-soprano, or soprano; change how fast the notes play in real time; slow down to focus on technique or speed up to improve agility; large display shows what note is playing so student can focus on passage areas or know when their range has increased; helpful tips and resources; designed for performing vocalists, voice students, karaoke enthusi-asts, singing teachers, choir directors, and music teachers; (iOS 4.0 or later).

SmithMusic 123 Voice Warm-Up (Robert Smith) Offers the convenience of a trainer, warm-up, and recorder; can be used for voice game challenges; student is the judge, coach, and trainer of their sound; for those that desire to sing and need a simple way of training the voice; for singers that know they have to warm up their voice to protect it from strain; for those that want to increase their range, improve the quality of their tone, and improve their ear; good for choirs, ensembles, quartets, duos, and soloists; records for playback; records any other musical parts or pieces and stores them in a library; (iOS 6.0 or later).

Vocal Warm Up by Musicopoulos (http://www.musicopoulos.com/vocal-warm-up-singers-iphone-ipad-app) Play the exercises in any order; create cus-tom playlists; track usage and progress; video demonstrations; tempo control; range control; (iOS 5.1 or later).

VocalEase (http://www.downrecs.com/) Portable warm-up studio for musi-cians and public speakers; designed by well-known vocal trainer and profes-sional vocalist Arnold McCuller; twelve warm-up exercises are included as selectable audio files to prepare the user for any type of vocal performance; exercises help any vocalist or public speaker warm up; also includes vocal instruction for females from legendary vocal coach Rosemary Butler; vocal exercises will automatically download the first time each is played; (iOS 4.3 or later; Android 1.6 and up).

Voice Builder (http://www.voicebuilderapp.com/welcome.html) App for bringing power, resonance, range, flexibility, and endurance to both the sing-ing and speaking voice; based on the Catona Voice Building System; vocal exercise program for developing the muscles of the human voice; created by renowned voice builder Gary Catona; (iOS 4.3 or later).

Voice Vocal Exercises (http://patrickqkelly.com/index.php/universal/vocal-exercises) Warm-ups with piano accompaniment; suggestions for each exer-cise; pick a starting note from the keyboard or the staff in treble or bass clef; the top note of the exercise is displayed in orange; transpose up or down a step when the exercise repeats; adjust the tempo; (iOS 5.0 or later).

· *6* ·

Brass Instruments Apps

ALL BRASS INSTRUMENTS APPS

Brass and Woodwind Instrument Sounds (Lesson Portal, LLC) Fun and educational app for young children ages two to five years; tap and hear brass and woodwind musical sounds; designed by a music teacher and tested by a two-year-old; launch the app and tap the picture of the instrument to hear; each sound is professionally recorded in full digital quality from real musicians and instruments; sounds can be overlapped or tapped to hear again; includes flute, clarinet, saxophone, trumpet, French horn, trombone, and tuba; (iOS 2.2.1 or later).

Brass Instrument Guides iPad Apps (http://www.iphone-ipad-iapp.com/ipad.html) Separate apps include: Trumpet Guide, Trombone Guide, Horn Guide, Euphonium Guide, and Tuba Guide; each app includes a method for working the musical instrument including long tones, mouthpiece buzzing, lip slurs, and tonguing; allows the user to record exercises; contains a tuning fork (B♭ 3, 440), advice on each lesson, and a bibliography on the instrument; the warm-up is the most important part of a practice session for developing endurance, muscle building, tone quality, intonation, and technique; (iOS 6.0 or later).

Brass InstrumentSS and ***Brass InstrumentSS IA*** (http://hdoapp.sakura.ne.jp) Music app with multiple brass instruments; includes Piccolo TrumpetSS, TrumpetSS, TromboneSS, French HornSS, TubaSS, Bass TrumpetSS, Bass TromboneSS, FlugelhornSS, CornetSS, AlphornSS, EuphoniumSS, CimbassoSS, SousaphoneSS, and WagnerTubaSS; easily play the trumpet with a single finger; will be able to play by touching the inner side of the white dotted line; volume slider at the bottom center of the screen; recording function; send recorded data via e-mail; supports iTunes File Sharing; play along with favorite songs; select the audio file format when recording; display scale on keyboard; support orientation; record/play/stop buttons; shows instrument ranges; (iOS 7.0 or later).

BrassNotes (http://www.universeapps.com/index.php/2-music-apps) Check the fingering or position of a note while playing or practicing; instruments included are trombone, euphonium, trumpet, and tuba; good for other instruments to transpose music; includes both treble clef and bass clef; stylized dark backgrounds; Universal app; Retina display; multitasking supported; e-mail support; (iOS 4.3 or later).

Fingering Brass (http://patrickqkelly.com) Comprehensive fingering charts for brass instruments; pick a written note by touching the staff and have the fingering displayed and concert pitch played on the piano; play a concert pitch on the piano and have the fingering displayed and the transposed note written on the staff; provides fingering charts, including many alternate fingerings, for French horn, trumpets including cornet and flugelhorn, trombones, euphonium and baritone, tubas, and a piano mode that displays the note name, location on the piano, and the notation in four different clefs: treble, alto, tenor, and bass; plays the concert pitch for any written pitch; sounds are included from C0 (below) to C9 (above), outside the range of the piano; Middle C is C4; treble clef on brass instruments follows the British Band tradition of being in transposing notation for non-transposing instruments; affects trombone, euphonium, baritone, and B♭, E♭, and F tubas; touch and drag up and down on the staff to select the note, slide right for sharp and left for flat, or slide up and down near either edge for constant sharps or flats; when playing the keyboard, swipe to move the keyboard, tap to play notes, touch and hold then slide to glissando; glissando up, notes will be notated in sharps; glissando down, notes will be notated in flats; if there are alternate fingerings available for a note, a button will appear allowing user to navigate through them; (iOS 7.0 or later).

Old Instruments Apps (http://www.iphone-ipad-iapp.com/iphone.html) Old Brass: app includes tablature of former instruments the Serpent and the Ophicleide; Baroque Trumpet: app includes the main fingerings for the baroque trumpet; Retina display compatible; (iOS 7.0 or later).

FRENCH HORN APPS

French Horn Pro HD (http://www.contactplus.com/frenchhornpro/) Learn how to play and master the French horn (F, B♭, and double horn); designed for real French horn players who want to improve ear training, fingering technique, accuracy, and speed; beginners who are just learning the French horn fingerings can slow the tempo down and watch as the correct fingerings appear on the screen; music practice is about repetition, muscle memory, and ear training; can change the fingerings to accommodate horn; videos and tutorials; real music staff is displayed and highlighted so user can follow the notes on the screen when learning a new song; can hear the song being played in real time; tells user immediately if they've hit the correct notes or made a mistake; tap a button to go to another song; track progress by keeping a cumulative score for each song, the total elapsed time spent practicing each song, remembering the practice tempo for each song, and comparing scores to user's friends; transpose notes while practicing by swiping finger up or down; can change the tempo by dragging fingertips across the staff; score will increase as user accurately presses the correct buttons at higher tempos; includes more than six hundred built-in songs including scales, exercises, and finger twisters; all included songs can be modified; includes ninety-nine blank templates; (iOS 4.0 or later).

PlayAlong French Horn (http://www.atplaymusic.com/our-apps/playalong-french-horn/) Listens to playing and guides user through the selected song; choose the settings that complement skills; cumulative statistics of the total correct notes, consecutive correct notes, and the songs played correctly are available; beginners can choose to have finger charts displayed under each note on the staff as they play; touching the finger chart lets the user hear that note; pushing the play button allows the user to hear the whole song; advanced users can choose to display only the note names under the staff or to see a full screen of music; in Manual mode the user can play the song with their own interpretation, using the display like traditional sheet music; Stats display presents a list of all the songs, showing which have been played and how many of the notes the user was able to play; History feature keeps track of scores for past performances for each song played; Transposition feature lets the user shift the pitch of a song's notes up or down to customize the range of the song; in Game Center stats are listed on the leader boards and songs are presented as accomplishments; includes some free songs and a chromatic scale; song list includes ratings of easy, intermediate, and advanced; share by e-mailing performances; (iOS 5.1 or later).

TROMBONE APPS

iBone (http://ibone.spoonjack.com) Practice and learn scales and songs; play along with music from own library; accompany the lead or take the solo spot; play along with a tune from the iBone Songbook; touch or blow to make a sound; slide finger to change pitch; raise and lower the bell to change volume; can cover the trombone's standard two-and-a-half-octave range, slur up and down, and slide in and out; play along with the band accompaniment; tap the rings at the bottom of their fall to hit the right note at the right time; comes with seven free songs; additional songs can be purchased and downloaded; (iOS 3.1 or later; Android 2.1 and up).

PlayAlong Trombone (http://www.atplaymusic.com/our-apps/playalong-trombone/) Listens to playing and guides user through the selected song;

choose the settings that complement skills; cumulative statistics of the total correct notes, consecutive correct notes, and the songs played correctly are available; beginners can choose to have finger charts displayed under each note on the staff as they play; touching the finger chart lets the user hear that note; pushing the play button allows the user to hear the whole song; advanced users can choose to display only the note names under the staff or to see a full screen of music; in Manual mode the user can play the song with their own interpretation, using the display like traditional sheet music; Stats display presents a list of all the songs, showing which have been played and how many of the notes the user was able to play; History feature keeps track of scores for past performances for each song played; Transposition feature lets the user shift the pitch of a song's notes up or down to customize the range of the song; in Game Center stats are listed on the leader boards and songs are presented as accomplishments; includes some free songs and a chromatic scale; song list ratings of easy, intermediate, and advanced; share by e-mailing performances; (iOS 5.1 or later).

Trombone musicofx (http://www.musicofx.com/trombone.php) Operates under the same principles as a trombone; real trombonists tighten or loosen their lips to select a partial including B♭, F, the B♭ above that, etc., according to the overtone series and can use the slide to change the note; the farther out they push the slide, the lower the resulting pitch; the slide lets a trombonist get at the notes between the partials; doesn't have a brass mouthpiece; can pick how to simulate the mouthpiece either by tilt, breath, or touch; touch the slide to move it out; "Snap to Slide Position" option that will put frets down on the slide; stay fretless to do glissandos; in addition to real trombone samples, app includes the Visual Synth; can combine waves of various shapes to create complex timbres; can sculpt the overtones and see the resulting wave; apply the low-pass filter to take the edge off and some reverb to add a little echo; (iOS 2.2 or later).

Trombone Pro HD (http://www.contactplus.com/trombonepro) For trombone players, beginners and advanced; will play exercises in bass clef and enable user to hear the chords and music and play along by dragging finger on slide; video tutorial; thirty-two measures of notes; can edit the music on screen; comes with more than five hundred exercises and gives user the ability to share exercises with other users; (iOS 5.0 or later).

TRUMPET APPS

iTrump (http://itrump.spoonjack.com) Practice and learn trumpet scales and songs; jam along with anything in iPod; play along with a song in the Songbook; useful to students and playable by anyone; touch or blow to make a sound; tap the "valve pads" to set the pitch; raise and lower the bell for volume; covers the trumpet's standard notated range; slur up and down, trill, bend notes, and add vibrato; pick a tune from the Songbook and app shows user how to play along with the band accompaniment; tap the colored blocks at the bottom of their fall to hit the right note at the right time; comes with seven songs; (iOS 3.1 or later).

Little Composers Trumpet 3 (http://littlecomposers.com) Features three modules designed to learn, play, and compose music; Module 1 is about learning and includes lessons; Module 2 is about playing songs; Module 3 lets children create their own lessons and songs; video tutorials; web forum; (iOS 3.2 or later).

Little Trumpet HD (http://littlecomposers.com) Learning experience for little trumpet players; eight colored valves; six additional hidden valves for more advanced players; built-in lessons; touch-sensitive notes that play every melody; built-in children's songs; ear training; discussion forum; (iOS 4.0 or later).

PlayAlong Trumpet (http://www.atplaymusic.com/our-apps/playalong-trumpet/) Listens to playing and guides user through the selected song; choose the settings that complement skills; cumulative statistics of the total correct notes, consecutive correct notes, and the songs played correctly are available; beginners can choose to have finger charts displayed under each note on the staff as they play; touching the finger chart lets the user hear that note; pushing the play button allows the user to hear the whole song; advanced users can choose to display only the note names under the staff or to see a full screen of music; in Manual mode the user can play the song with their own interpretation, using the display like traditional sheet music; Stats display presents a list of all the songs, showing which have been played and how many of the notes the user was able to play; History feature keeps track of scores for past performances for each song played; Transposition feature lets the user shift the pitch of a song's notes up or down to customize the range of the song; in Game Center stats are listed on the leader boards and songs are presented as accomplishments; includes free songs and a chromatic scale; song list ratings of easy, intermediate, and advanced; share by e-mailing performances; (iOS 5.1 or later).

Trumpet musicofx (http://www.musicofx.com/trumpet.php) App is a trumpet in the iPhone; trumpeters tighten or loosen their lips to select a partial including B♭, F, the B♭ above that, etc., according to the overtone series; can press the valves to change the note; the more valves depressed, the lower the resulting pitch; valves let a trumpeter get at the notes between the partials; doesn't have a brass mouthpiece; can pick how to stimulate the mouthpiece either by tilt, breath, or touch; depress the valves to change the pitch; manual includes a fingering chart; app includes a realistic partial option that enables faster, more natural playing; learn the principles of the trumpet and hone skills; in addition to a sampled trumpet and mute samples, app includes the Visual Synth; can combine waves of various shapes to create complex timbres; sculpt the overtones and see the resulting wave; (iOS 2.2 or later).

Trumpet Pro HD (http://www.contactplus.com/trumpetpro/) Practice trumpet fingerings with ear training; includes thirty-two measures per song; can modify existing songs and add new songs; video tutorial; experience ear training by hearing the chords and sounds corresponding to the notes; includes C/B♭ trumpet pitch; 986 possible chords; large song library; song editing with copy/paste between songs; share songs via e-mail or post on a website or blog; tempo adjustment per song or global; transpose up/down while practicing; count-in before song starts; scoring for accuracy; wrong notes are highlighted;

music practice is about repetition, muscle memory, and ear training; achieve trumpet playing goals; (iOS 5.0 or later).

Trumpet Studio (Windy Town) For every level of trumpet player; interactive, animated fingering chart with sound; press the arrow up or down to see the fingering; touch the speaker to hear the pitch; articles written in clear, concise, and easy-to-understand language; articles on: High Note Playing, Air Exercises and Aids, Method Book List (easy, medium, difficult), Mouthpiece Advice, Lip Buzzing Exercises, and Practice Routines; digital documents including: Blues Charts, High Note/High Range with Exercises, Scales including Major, Minor, Melodic, Harmonic, Technical Study Exercises, Trumpet Staff with Piano Part, and Brass Quintet/Quartet Staff; digital files can be downloaded and printed for use and are included in the app as a reference; (iOS 2.2.1 or later).

Virtual Trumpet (https://sites.google.com/site/snowysapps/virtual-trumpet) Play the trumpet on smartphone or tablet; can download other instruments available on Google Play such as guitar, piano, and drums and play together with friends; sound quality may vary depending on the device; (Android 2.0 and up).

TUBA APPS

PlayAlong Tuba (http://www.atplaymusic.com/our-apps/playalong-tuba/) Listens to playing and guides user through the selected song; choose the settings that complement skills; cumulative statistics of the total correct notes, consecutive correct notes, and the songs played correctly are available; beginners can choose to have finger charts displayed under each note on the staff as they play; touching the finger chart lets the user hear that note; pushing the play button allows the user to hear the whole song; advanced users can choose to display only the note names under the staff or to see a full screen of music; in Manual mode the user can play the song with their own interpretation, using the display like traditional sheet music; Stats display presents a list of all the songs, showing which have been played and how many of the notes the user was able to play; History feature keeps track of scores for past performances for each song played; Transposition feature lets the user shift the pitch of a song's notes up or down to customize the range of the song; in Game Center stats are listed on the leader boards and songs are presented as accomplishments; includes free songs and a chromatic scale; song list includes ratings of easy, intermediate, and advanced; share by e-mailing performances; (iOS 5.1 or later).

Tuba—Amazing Videos (Mobile Apps Services) Tuba tutorials and video lessons for beginners or advanced players; videos are chosen from YouTube and presented as is; watch the best tuba players in the world; (Android 2.2 and up).

Tuba Notes Flash Cards (http://aptapp.blogspot.com) Flashcards to develop note reading on the bass clef for all tuba players in any style of music; front side of the card displays the note on the music staff; back side displays what note it is and the fingering; (Android 2.2 and up).

· 7 ·

Woodwind Instruments Apps

ALL WOODWIND INSTRUMENTS APPS

Articulations (http://www.iphone-ipad-iapp.com/ipad.html) Method for working dynamics, tenuto, staccato, accent, double tongue, and triple tongue on the reed instruments including bassoon, oboe, clarinet, and saxophone; basis for tackling difficult passages in the classic, modern, and jazz repertoire; advice on how to work each lesson, a bibliography on the subject, a list of articulations, and a tuning fork (A3, 440); (iOS 7.0 or later).

Fingering for iPad (http://patrickqkelly.com) A fingering chart is an essential reference tool; includes more than 1,100 fingerings for woodwind and brass instruments; includes alternative fingerings and trill fingerings; can be used to learn other instruments; (iOS 7.0 or later).

FingeringCharts Pro (http://scottrundell.com/FingeringChart.html) Provides pitches and fingerings for twenty-one instruments; tap and drag for fine control; for all levels of students and educators; easy interface with fingerings and pitches for new instruments in one place; transposed pitch reference; swipe up and down to change notes; tap and drag for better control; tap the play button to play the pitch; includes basic and up to four alternate fingerings for Piccolo, flute, clarinet, bass clarinet, oboe, soprano saxophone, alto saxophone, tenor saxophone, baritone sax, bassoon, trumpet, single French horn, double French horn, trombone (no trigger), trombone (F trigger), bass trombone, baritone/euphonium 4 valve, baritone/euphonium 3 valve, baritone/euphonium 3 valve TC, tuba (3 valve), and tuba (4 valve); bass and treble clefs; option to change number of fingerings shown; plays the pitch shown; (iOS 5.0 or later).

Fingering Woodwinds (http://patrickqkelly.com) Comprehensive fingering charts for woodwind instruments; pick a written note and have the fingering displayed and concert pitch played; a touch of a button reveals the note on the piano; play a concert pitch on the piano and have that pitch's fingering displayed; touch of a button reveals the transposed note written on the staff; more than 1,200 fingerings, including many alternates for the following instruments: flutes (piccolo, concert, alto, and bass) including m2, M2, m3, M3 trill fingerings; oboe and cor anglais including m2 and M2 trill fingerings; clarinets (soprano, alto, bass, and contrabass) including m2, M2 trill fingerings; bassoon and contrabassoon including m2 and M2 trill fingerings (treble, tenor, and bass clefs); saxophones (soprano, alto, tenor, and baritone) includes m2, M2, m3, M3 trill fingerings; alto and tenor Sax include altissimo fingerings up to written F, two octaves above the top line F of the treble clef; piano mode displays the note name, location on the piano, and the notation in four different clefs: treble, alto, tenor, and bass; plays the concert pitch for any written pitch; sounds are included from C0 (below) to C9 (above), outside the range of the piano; includes visual guides to the woodwind key names and locations to understand the fingering charts better; play the trills for trill fingerings; alto and tenor saxophone altissimo fingerings are supplied by Steve Moran; touch and drag up and down on the staff to select the note, slide right for sharp and left for flat, or slide up and down near either edge for constant sharps or flats; when playing the keyboard, swipe to move the keyboard, tap to play notes, touch and hold, then slide to glissando; glissando up, notes will be notated in sharps; glissando down, notes will be notated in flats; when tapping specific notes, they will be notated in the most common accidental for that note; if there are alternate fingerings available for a note or trill, a button allows the user to navigate through them; includes a mute button as well as a step controller to move chromatically up and down without touching the music staff or piano keyboard; (iOS 7.0 or later).

Old Instruments Apps by Ambroise Charron (http://www.iphone-ipad-iapp .com/iphone.html) Old Bassoon: app includes tablature of these former bassoons: Bünher & Keller, Eichentopf, Rottenburg, Grenser, Sherer, and romantic bassoon; Old Flute: app includes tablature of these former recorders: Virdung, Ganassi, Jambe de Fer, Mersenne, Bismantova, Agricola, Hotteterre, Freilhon-Poncein, Carr, and Blankenburg; Old Oboe: app includes tablature of these former oboes: Stanesby, Denner, Grassi, saxon oboe, and two keys oboe; Traverso: app includes the tablature of these Baroque flutes: the Renaissance flute, the Beukers flute, and the one key flute; Sarrusophone: app includes the tablature of the Sarrusophone; Dulcian: app groups the main fingerings for the soprano dulcian, the alto ducian, the tenor dulcian, and the bass dulcian; Shawm: app groups the main fingerings for the sopranino shawm, the soprano shawm, and the alto shawm; Crumhorn: app groups the main fingerings for the soprano crumhorn, the alto (in F) crumhorn, the alto (in G) crumhorn, the tenor crumhorn, and the bass crumhorn; Old Diapason: app includes the five most used tuning forks: A3 392 Hz, A3 415 Hz, A3 421 Hz, A3 430 Hz, and A3 440 Hz; all apps are Retina display compatible; (iOS, varies by app).

Reeds Scraping Apps (http://www.iphone-ipad-iapp.com/iphone.html) Bassoonreeds: app tells user the areas to be scraping on their bassoon reed to improve it; database for scraping reeds of the user; Oboereeds: app tells user the areas to be scraping on their oboe reed to improve it; database for scraping reeds of the user; Clarinetreeds: app tells user the areas to be scraping on their clarinet reed to improve it; database for scraping reeds of the user; Saxoreeds: app tells user the areas to be scraping on their saxophone reed to improve it; database for scraping reeds of the user; all apps are Retina display compatible and Universal; (iOS 7.0 or later).

Scales Book (http://www.iphone-ipad-iapp.com/ipad.html) Method for working scales, arpeggios, and chords on the woodwinds and brasswinds including flute, bassoon, oboe, recorder, clarinet, saxophone, trumpet, trombone, euphonium, tuba, and French horn; contains scales including major, minors, be-bop, blues scales, modal scales, harmonic scales, pentatonic, altered pentatonic, modes of limited transposition, intervals, chromaticism, arpeggios, chords with extensions, and a list of articulations; basis for tackling difficult passages in the classic, modern, and jazz repertoire; includes advice and a bibliography; (iOS 7.0 or later).

Woodwind Fingering Chart (http://adamfoster.net/blog/) Fingering chart for woodwind instruments; includes flute and clarinet; press menu to change instruments; swipe up or down to change notes; swipe left or right to change fingering; drop-downs at bottom of screen can be used to change note types including single notes, trills, tremolos, etc.; notes and fingerings; red marks are keys to press down, green marks are keys to trill; drag left and right in the title bar to see longer note descriptions; (Android 2.0.1 and up).

Woodwind InstrumentSS IA (http://hdoapp.sakura.ne.jp); Music application with multiple woodwind instruments; allows the user to play the instrument with a single finger; will be able to play by touching the inner side of the white dotted line; includes English HornSS, FluteSS, ClarinetSS, OboeSS, BassoonSS, Tenor SaxophoneSS, PiccoloSS, OcarinaSS, RecorderSS, Pan-FluteSS, Alto SaxophoneSS, QuenaSS, Bass ClarinetSS, and ZamponasSS; (iOS 7.0 or later).

BASSOON APPS

Bassoon (http://apehq.com) Includes basic bassoon fingering chart with additional content; (Android 1.5 and up).

Bassoon Fingerings (Mike Muszynski) Allows for easy retrieval of standard fingerings as well as the ability to save custom personal fingerings and notations; preloaded with standard fingerings compiled over the developer's fifteen years of study; fingerings are displayed using standard notation with holes representing the major finger placements and keys based on standard key names; fingerings can be saved for notes as high as high F and as low as low B♭; for any serious bassoon player, student, teacher, or band director; (iOS 2.0.1 or later).

Bassoon Flash Cards (Christian Liang) Flashcards to develop note reading on the bass clef for bassoon players in any style of music; front side of the card

displays the note on the music staff; back side displays finger position; audio reference for each note; (iOS 1.0 or later).

Bassoon Guide (http://www.iphone-ipad-iapp.com/ipad.html) Method for working the instrument including long tones, embouchure flexibility, finger technique, half-hole technique, chromaticism, and the upper register; allows the user to record exercises while working to ensure intonation and tone homogeneity; basis for tackling difficult passages in the bassoon repertoire; tuning fork (A2, 440); lesson advice; bibliography on the bassoon; (iOS 6.0 or later).

Contrabassoon (http://www.iphone-ipad-iapp.com/iphone.html) Shows contrabassoon tablature, including German and French systems; Retina display compatible; (iOS 7.0 or later).

CLARINET APPS

Clarinet Fingering Chart (FuFio Cat) Learn to play the clarinet; simple and easy to use; shows all the common fingering and trill combinations available on the clarinet; play the note to control if execution is correct (B♭ clarinet); note played is a real sound from a professional clarinetist; (Android 2.2 and up).

Clarinet Fingering Chart (http://www.bearcatbands.com) Easy-to-use fingering reference for the modern clarinet; covers over a three-octave range; includes alternate fingerings when appropriate; tap the button of the note and the fingering as well as the printed note will appear; organized by register name: chalumeau, throat tones, clarion, and altissimo (up to G); helps young students learn the proper names of the registers; all keys are included on each diagram to know precisely where to put fingers; buttons are laid out in an enharmonic fashion; reference for both students and teachers; (iOS 6.1 or later).

Clarinet Fingering Chart (http://www.critical-apps.co.uk) Reference application that shows all the common fingering and trill combinations available on the clarinet including more than four hundred different combinations from E3 to A7; scroll left or right along the staff to the note and tap to select; tap the note again or use the back button to return to the staff; for fingerings with additional information, tap the "I" button to show more info; to hear the note being played by the instrument, press the "Play Note" button; (Android, varies with device).

Clarinet Guide (http://www.iphone-ipad-iapp.com/ipad.html) Method for working the instrument including long tones, embouchure flexibility, finger technique, chalumeau to clarion, chromaticism, and the altissimo register; allows the user to record exercises while working to ensure intonation and tone homogeneity; basis for tackling difficult passages in the clarinet repertoire; contains a tuning fork (A3, 440), advice on each lesson, and a bibliography on the clarinet; (iOS 6.0 or later).

Clarinet in Reach (http://musicinreach.com) By Metropolitan Opera Orchestra clarinetist Anthony McGill; designed for the musical education of both beginners as well as advancing players; standard fingering chart as well as trill fingering chart; musical term guide; audio files of many popular études for clarinet players; video help guides to assist in playing development; equipment guide for beginning and professional players; (iOS 4.0 or later).

Clarinet musicofx (http://www.musicofx.com/clarinet.php) Play the clarinet on the iPhone; almost every key on the Boehm-system standard clarinet is included and appropriately animated, including all the pinky keys; go from low E to a high E; there's no octave key on the back of the iPhone so instead the user can tilt the iPhone up; use breath control; sampling lets user articulate as fast as on a real clarinet; hold it away and play via touch; learn the principles of the clarinet; manual includes a fingering chart; some notes have multiple fingerings; in addition to a sampled B♭, A, E♭, and bass clarinet, app includes the Visual Synth; combine waves of various shapes to create complex timbres; sculpt the overtones and see the resulting wave; apply the low-pass filter to take the edge off and some reverb to add a little echo; (iOS 2.2.1 or later).

Easy Clarinet—Clarinet Tuner (Games and Study Tools) Tune clarinet, saxophone, trumpet, French horn, trombone, tuba, or any brass instrument quickly and efficiently; for marching band or musical lessons or practice; brass tuner uses a combination of probability theory and Fourier transforms to provide the best chromatic tuner for string, woodwind, or brass instruments; aimed at those who play the clarinet; includes sample notes for Concert B♭, Concert F, and Concert A440 for speed and accuracy for tuning all band and jazz instruments; by using the guiding notes plus the auto-tuner functionality user can quickly tune any woodwind instrument with this clarinet tuner; (Android 2.2 and up).

How to Play Clarinet (Josef Pavlik) Clarinet fingering reference chart with more than 140 different fingerings from E3 to A7; press the piano key to hear the clarinet sound and see the fingering; can choose one of the eleven different types of clarinet from contra-bass clarinet (B♭) to piccolo clarinet (A♭); play the sampled sound on Android while playing own clarinet to compare the pitch; can play the app like a piano or like a harmonica; (Android 2.1 and up).

PlayAlong Clarinet (http://www.atplaymusic.com/our-apps/playalong-clarinet/) Listens to playing and guides user through the selected song; choose the settings that complement skills; cumulative statistics of the total correct notes, consecutive correct notes, and the songs played correctly are available; beginners can choose to have finger charts displayed under each note on the staff as they play; touching the finger chart lets the user hear that note; pushing the play button allows the user to hear the whole song; advanced users can choose to display only the note names under the staff or to see a full screen of music; in Manual mode the user can play the song with their own interpretation using the display like traditional sheet music; Stats display presents a list of all the songs, showing which have been played and how many of the notes the user was able to play; History feature keeps track of scores for past performances for each song played; Transposition feature lets the user shift the pitch of a song's notes up or down to customize the range of the song; in Game Center stats are listed on the leader boards and songs are presented as accomplishments; includes some free songs and a chromatic scale; song list ratings of easy, intermediate, and advanced; share by e-mailing performances; (iOS 5.1 or later).

Real Clarinet (Mauro Apps) Created by a music teacher; portable and functional; suitable for students and those who want to play; has all the clarinet notes; the clarinet is a type of woodwind instrument that has a single-reed mouthpiece, a straight cylindrical tube with an approximately cylindrical bore, and a flaring bell; there are many types of clarinets with differing sizes and pitches, comprising a large family of instruments; the unmodified word "clarinet" usually refers to the B♭ soprano clarinet, the most common type, with a large range of nearly four octaves; the clarinet family is the largest woodwind family, with more than a dozen types, ranging from the extremely rare BBB♭ octo-contrabass to the A♭ piccolo clarinet; of these, many are rare or obsolete, and music written for them is usually played on more common versions of the instrument; Johann Christoph Denner invented the clarinet in Germany around the turn of the eighteenth century by adding a register key to the earlier chalumeau; additional keywork and airtight pads were added to improve tone and playability; the clarinet is used in jazz and classical ensembles, in chamber groups, and as a solo instrument; (Android 2.2 and up).

Real Clarinet (Volkeman) Sensitive to breath; easy to learn; blow and hear the clarinet; see all combinations in the settings; (iOS 4.0 or later).

FLUTE AND PICCOLO APPS

Bamboo Flute Pro (http://agileant.tumblr.com) Sensitive to breath, touch, and movements; touch combinations of holes to play notes; adjust the inclination to change the octave; up to nineteen individual real bamboo flute notes sampling; full reverb effect; (iOS 4.0 or later).

Flute (http://rskapps.eu5.org/androidapps.php) Press fingers on the "holes" to play melodies; designed with real flute sounds; (Android 2.2 and up).

Flute Contemporary (http://www.iphone-ipad-iapp.com/ipad.html) Method for working extended technique of the flute including flutter-tongue, whistle-tone, aeolian sounds, tongue-ram, pizzicato, and key percussion; allows the user to record exercises while working to ensure intonation and tone homogeneity; basis for tackling difficult passages in the modern flute repertoire and jazz music; advice on each lesson; bibliography on the subject; (iOS 6.0 or later).

Flute Fingering Chart (http://www.critical-apps.co.uk) Reference application that shows all the common fingering and trill combinations available on the flute including more than four hundred different combinations for flute, piccolo, soprano, alto, and bass flutes, from B3 to F♯7; scroll left or right along the staff to the note and tap to select; tap the note again or use the back button to return to the staff; for fingerings with additional information, tap the "I" button to show more info; to hear the note being played, press the Play Note button; (Android, varies with device).

Flute Fingering Chart (http://www.fluteinfo.com/Fingering_chart/index.html) Created by the flautist at FluteInfo.com; educational tool that includes four octaves of standard and alternate fingerings for the flute and piccolo; includes all fingerings from B4 up to G7; contains trill fingerings for the first three octaves; (iOS 7.0 or later).

Flute Guide (http://www.iphone-ipad-iapp.com/ipad.html) Method for working the flute including long tones, harmonics, finger technique, medium register, chromaticism, and the upper register; allows the user to record exercises while working to ensure intonation and tone homogeneity; basis for tackling difficult passages in the flute repertoire; contains a tuning fork (A3, 440), advice on each lesson, and a bibliography on the flute; (iOS 6.0 or later).

Flute Intonation (http://www.iphone-ipad-iapp.com/iphone.html) Groups the main pitch alteration fingerings for the intonation of the flute; Retina display compatible; Universal app; (iOS 7.0 or later).

Flute Master (http://www.createfreeiphoneapps.com) Put mouth up to the screen and gently rock to activate the flute app; finely calibrated program will correspond to every rhythm; authentic flute sound; for beginners and advanced flautists; (iOS 2.2.1 or later).

Flute musicofx (http://www.musicofx.com/flute.php) App is an open-hole flute on the iPhone; plays and sounds like a flute; can adjust the octave breaks according to how user holds the flute and tilt to adjust breaks; use breath control and sampling for double and triple tonguing; demonstrate fingerings; learn to play the flute; includes a sampled alto flute, bass flute, shakuhachi and standard flute, and Visual Synth; combine waves of various shapes to create complex timbres; sculpt the overtones; apply the low-pass filter to take the edge off and some reverb to add a little echo; (iOS 2.2.1 or later).

Flute Scales (for Flute and Piccolo) (TheWay, Ltd.) Explore the flute or piccolo; includes scales from around the world displayed in a manner easy for beginners and experts; displays scales and finger diagrams on a single page, covering the range of the flute from fourth octave C to seventh octave F; basic fingering diagrams for the modern Boehm system flute and piccolo (flute key of C); real flute sounds; touch any fingering diagram to play the associated note; interval values shown next to each note in scale; all twelve keys for all scales; sixty-three scales included; (iOS 5.1 or later).

Flute Study (http://www.iphone-ipad-iapp.com/ipad.html) Method in twelve lessons for learning to play the flute; exercises for developing endurance, muscle building, tone quality, intonation, technique, and musicality; allows the user to record exercises while working to ensure intonation and tone homogeneity; basis for playing the flute repertoire; contains a full tab of the flute, a tuning fork (A3, 440), a lexicon, and advice on how to work each lesson; (iOS 6.0 or later).

Flute Timbre (http://www.iphone-ipad-iapp.com/iphone.html) App includes the fingerings for producing variations in tone color on the flute; Retina display; Universal app; (iOS 7.0 or later).

Irish Flute (http://www.tradlessons.com/IrishFlute.html) Authentic-sounding wooden flute for the iPad; for anyone interested in learning to play the Irish flute or pennywhistle or someone who already plays the Irish flute and wants to practice tunes on the iPad; fingerings are based on those used for the real instrument and familiar to Irish pennywhistle and flute players; sounds are from high-quality per-note recordings of an 1857 British Metzler wooden flute; demonstration video; holes light up when touched; use the fingering chart as a guide; amount of vibrato is adjustable; all settings are saved and restored the next time the app is run; (iOS 5.1.or later).

Magic Flute (http://www.powermme.com/magicflute/cn/) Music flute game with sound effects; play the flute melody mixed with drum, piano, guitar, and other instruments; includes more than one hundred songs; solo play the flute and save performances; (Android 2.2 and up).

Magic Flute for Little Composers (http://littlecomposers.com/) For beginners; colorful play pads; built in lessons; real songs to learn; ear training; jam-along sessions for advanced students; unique jam module for right- and left-handed children; introductory video; (iOS 3.2 or later).

Ocarina (http://www.smule.com/apps#ocarina) Turns iPhone into an ancient flute; sensitive to breath, touch, and movements; more versatile than an actual flute; blow air into mic to create music; touch combinations of holes to play notes; tilt to change vibrato rate and depth; advanced users can change keys and modes, including C Major and Zeldarian mode; hear people learning to play around the world by tapping the globe icon; rate and share favorite performances via e-mail; record and archive favorite performances on "My Ocarina" web page; (iOS 4.0 or later).

Piccolo Alternate (http://www.iphone-ipad-iapp.com/iphone.html) Groups the main pitch alteration fingerings for the intonation of the piccolo and the main alternate fingerings; Retina display; Universal app; (iOS 6.0 or later).

Piccolo Harmonics (http://www.iphone-ipad-iapp.com/ipad.html) A piccolo tube can produce the series of natural harmonics of a note by changing the pressure of the lips and the air column without changing the fingerings; playing harmonics will enable the piccolo player to gain a more solid foundation in their normal tone and help them acquire the required control and flexibility to perform the upper register; exercises for developing the muscles of the lips and the ear; allows the user to record exercises while working to ensure intonation and tone homogeneity; basis for playing the upper register and developing a personal sound; (iOS 7.0 or later).

PlayAlong Flute (http://www.atplaymusic.com/our-apps/playalong-flute/) Listens to playing and guides user through the selected song; choose the settings that complement skills; cumulative statistics of the total correct notes, consecutive correct notes, and the songs played correctly are available; beginners can choose to have finger charts displayed under each note on the staff as they play; touching the finger chart lets the user hear that note; pushing the play button allows the user to hear the whole song; advanced users can choose to display only the note names under the staff or to see a full screen of music; in Manual mode the user can play the song with their own interpretation, using the display like traditional sheet music; Stats display presents a list of all the songs, showing which have been played and how many of the notes the user was able to play; History feature keeps track of scores for past performances for each song played; Transposition feature lets the user shift the pitch of a song's notes up or down to customize the range of the song; in Game Center stats are listed on the leader boards and songs are presented as accomplishments; includes free songs and a chromatic scale; song list includes ratings of easy, intermediate, and advanced; share by e-mailing performances; (iOS 5.1 or later).

Pocket Flute (http://www.createfreeiphoneapps.com) Put mouth up to screen and gently rock to activate the flute sound; responds to every rhythm, producing an authentic flute sound; for all flautists; (iOS 2.2.1 or later).

Tin Whistle (http://www.iphone-ipad-iapp.com/iphone.html) App includes the main fingerings for the Irish flute; Retina display; (iOS 7.0 or later).

Tin Whistle Study (http://www.iphone-ipad-iapp.com/ipad.html) Seven lessons to start learning to play the tin whistle; includes exercises for developing endurance, muscle building, tone quality, intonation, technique, and musicality; allows the user to record exercises while working to ensure intonation and tone homogeneity; basis for playing the tin whistle repertoire; contains a full tab, a tuning fork (A3, 440), a lexicon, and advice on each lesson; (iOS 6.0 or later).

OBOE AND ENGLISH HORN APPS

English Horn Flash Cards (http://www.classclef.com/app) The English horn is a large oboe pitched a fifth below the ordinary oboe; features flashcards to develop note reading on the treble clef for all English horn players in any style of music; front side of the card displays the note on the music staff; back side displays what note it is and what fingering to use; audio reference for each note; for development in sight-reading and audio recognition; (iOS 4.0 or later).

Oboe Alternate (http://www.iphone-ipad-iapp.com/iphone.html) Includes main harmonic fingerings, alternate fingerings, and microtonal fingerings; Retina display; (iOS 7.0 or later).

Oboe Fingering Chart (http://fox-custom-software.com) App includes a reference for common oboe fingerings; (Android 1.1 and up).

Oboe Guide (http://www.iphone-ipad-iapp.com/ipad.html) Method for working the instrument including long tones, embouchure flexibility, finger technique, medium register, chromaticism, and the upper register; allows the user to record exercises while working to ensure intonation and tone homogeneity; basis for tackling difficult passages in the oboe repertoire; tuning fork (A3, 440); advice on how to work each lesson; bibliography on the oboe; (iOS 6.0 or later).

RECORDER APPS

JoyTunes Recorder Master (http://www.joytunes.com/ipad-home.php) App used by thousands of music teachers; activated by playing a real recorder; sounds user plays control the game; introduction to music; game is activated by a regular soprano recorder and by touch; play notes on the recorder to avoid obstacles, collect bonuses, and scare away evil birds; play songs while working on rhythm, correct and stable tone production, ear training, fingering technique, and more; (iOS 6.0 or later).

Learn and Play Recorder (http://www.musicplay.ca) Teaches beginners about the recorder, how to read music, and how to play the soprano recorder; app is based on the Recorder Resource Kit by Denise Gagné used by hundreds of thousands of children in the United States and Canada to learn to play the recorder; All about the Recorder: students learn about different sizes of recorders, composers who wrote for the recorder, and how to begin playing the recorder; How to Care for the Recorder: students learn how to care for and clean the recorder; How Notes are Named: students learn the names of the notes and where they are found on the musical staff; Counting Music: introduces students to the note values that will be used in the songs they will learn to play; Let's Begin: students are taught how to tongue the notes and how to blow the recorder; the first note B is introduced; Soprano Fingering: reference section for students to use when they need to learn the fingering for a note; The Songs: includes thirty-eight songs all with performance tracks and a recorder playing the melody; students can listen to the performance and check that their own performance is accurate; the songs begin with Just B, Just A, and then A and B; songs are introduced in a sequence that takes students from playing just one note to playing the entire C scale; songs include themes of works by famous composers; press the name to read about the composer; can hide the composer biographies; (iOS 4.0 or later).

Recorder Guide (http://www.iphone-ipad-iapp.com/ipad.html) Method for working the instrument; long tones, finger technique, chromaticism, low register, middle register, upper register, and flutter-tongue; record exercises while working to ensure intonation and tone homogeneity; basis for tackling difficult passages in the recorder repertoire; contains a tuning fork (A3, 440), advice on each lesson, and a bibliography on the recorder; (iOS 6.0 or later).

Recorder Study (http://www.iphone-ipad-iapp.com/ipad.html) Method in thirteen lessons to start learning to play the recorder; includes exercises for developing endurance, muscle building, tone quality, intonation, technique, and musicality; allows the user to record exercises while working to ensure intonation and tone homogeneity; basis for playing the recorder repertoire; contains a full tab of the recorder, a tuning fork (A3, 440), a lexicon, and advice on how to work each lesson; (iOS 6.0 or later).

SAXOPHONE APPS

iFretless Sax (http://ifretless.com) App for guitar, bass, and strings players who need realistic woodwind instrument sounds for recording or live performance; responds naturally to the finger movements strings players use for vibrato and sliding between pitches; gives a more believable and natural imitation of woodwind sounds than traditional keyboard-based synth interfaces; supports MIDI input from EWI MIDI controller devices; includes tenor sax, alto sax, bari sax, soprano sax, clarinet, bass clarinet, and synth lead; dynamic volume control based on touch force; MIDI in and out; Inter-App Audio connects with GarageBand, Beatmaker 2, Cubasis, Auria, and other recording apps; Audiobus; AudioCopy; (iOS 7.0 or later).

Learn to Play Saxophone (Rico7) App for saxophone players; watch videos and learn to play the saxophone; includes Tips and Tricks, Learn To Play Saxophone For Beginners, Learn To Play Saxophone For Alto, Learn To Play Saxophone For Intermediate, Learn To Play Saxophone For Advanced, Learn To Play Saxophone Tenor, and Learn To Play Saxophone Jazz; for sax players of all ages; (Android 2.1 and up).

PlayAlong Alto Sax (http://www.atplaymusic.com/our-apps/playalong-alto-sax/) Listens to playing and guides user through the selected song; choose the settings that complement skills; cumulative statistics of the total correct notes, consecutive correct notes, and the songs played correctly are available; beginners can choose to have finger charts displayed under each note on the staff as they play; touching the finger chart lets the user hear that note; pushing the play button allows the user to hear the whole song; advanced users can choose to display only the note names under the staff or to see a full screen of music; in Manual mode the user can play the song with their own interpretation, using the display like traditional sheet music; Stats display presents a list of all the songs, showing which have been played and how many of the notes the user was able to play; History feature keeps track of scores for past performances for each song played; Transposition feature lets the user shift the pitch of a song's notes up or down to customize the range of the song; in Game Center stats are listed on the leader boards and songs are presented as accomplishments; includes some free songs and a chromatic scale; song list ratings of easy, intermediate, and advanced; share by e-mailing performances; (iOS 5.1 or later).

PlayAlong Tenor Sax (http://www.atplaymusic.com/our-apps/playalong-tenor-sax/) Listens to playing and guides user through the selected song; choose the settings that complement skills; cumulative statistics of the total correct notes, consecutive correct notes, and the songs played correctly are available; beginners can choose to have finger charts displayed under each note on the staff as they play; touching the finger chart lets the user hear that note; pushing the play button allows the user to hear the whole song; advanced users can choose to display only the note names under the staff or to see a full screen of music; in Manual mode the user can play the song with their own interpretation, using the display like traditional sheet music; Stats display presents a list of all the songs, showing which have been played and how many of the notes the user was able to play; History feature keeps track of scores for past performances for each song played; Transposition feature lets the user shift the pitch of a song's notes up or down to customize the range of the song; in Game Center stats are listed on the leader boards and songs are presented as accomplishments; includes some free songs and a chromatic scale; song list ratings of easy, intermediate, and advanced; share by e-mailing performances; (iOS 5.1 or later).

Saxophone (Awesome Info Apps) Information about saxophones; latest updates, news, information, videos, photos, and events; continuous, up-to-date daily stream of news and information about the saxophone; read and watch exclusive saxophone articles, images, audio, and videos; the latest saxophone products and deals; share saxophone photos, events, and other news through Twitter, Facebook, and e-mail; (Android 2.2 and up).

Saxophone All-in-One (http://saxophonehk.weebly.com) Learn about playing the saxophone; key chart for fingering; saxophone family; saxophone scales; blues scales; transposition chart; (Android 2.2 and up).

Saxophone Apps by Ambroise Charron (http://www.iphone-ipad-iapp.com/iphone.html) Soprano Intonation: app groups the main pitch alteration fingerings for the intonation of the soprano saxophone; Alto Intonation: app groups the main pitch alteration fingerings for the intonation of the alto saxophone; Tenor Intonation: app groups the main pitch alteration fingerings for the intonation of the tenor saxophone; Baritone Intonation: app groups the main pitch alteration fingerings for the intonation of the baritone saxophone; Saxo Alternate: app shows user the main alternate fingerings on saxophone; SaxoMicrotonal: app shows user the main fingerings for quarter tones on saxophone; SaxoTimbre: app shows user the main fingerings for producing substantial variations in tone color on saxophone; Saxo High Tones: app includes the main altissimo fingerings for the sopranino, soprano, alto, tenor, and baritone saxophones; all apps are Retina display compatible and Universal; (iOS 7.0 or later).

Saxophone Apps for iPad by Ambroise Charron (http://www.iphone-ipad-iapp.com/ipad.html) SaxoStudy E♭: app for alto saxophone; method in thirteen lessons to start learning to play the saxophone; includes exercises for developing endurance, muscle building, tone quality, intonation, technique, and musicality; allows the user to record exercises while working to ensure intonation and tone homogeneity; contains a full tab of the saxophone, a tuning fork (A3, 440), a lexicon, and advice on each lesson; SaxoStudy B♭: app for tenor saxophone; method in thirteen lessons to start learning to play the saxophone; includes exercises for developing endurance, muscle building, tone quality, intonation, technique, and musicality; allows the user to record exercises; contains a full tab of the saxophone, a tuning fork (A3, 440), a lexicon, and advice on each lesson; Altissimo E♭: app for alto saxophone; saxophone method to develop the altissimo register and add an octave to the usual range of the instrument; allows the user to record exercises; contains a full tab of the saxophone, fingerings of altissimo range, advice on each lesson, and a bibliography on the subject; Altissimo B♭: app for tenor saxophone; saxophone method to develop the altissimo register and add an octave to the usual range of the instrument; allows the user to record exercises; contains a full tab of the saxophone, fingerings of altissimo range, advice on each lesson, and a bibliography on the subject; Saxoeffects: method for working extended technique of the saxophone including overtones, slap-tongue, subtone, flutter-tongue, growl, vibrato, and bending; allows the user to record exercises; contains a full tab of the saxophone, fingerings of altissimo range, advice on each lesson, and a bibliography on the subject; Saxotechnic: method for working the saxophone including finger technique, chromaticism, upper register, low register, hightones register, and alternate fingerings; allows the user to record exercises; contains a full tab of the saxophone, fingerings of altissimo range, advice on each lesson, and a bibliography on the subject; Saxosound E♭: app for alto saxophone; method to develop a personal sound; allows the user to record exercises; contains a full tab of the saxophone, fingerings of altissimo range, advice on each lesson, and a bibliography on the subject; Saxosound B♭: app for tenor saxophone; method to develop a personal sound; allows the user to record exercises; contains a full tab of the saxophone, fingerings of altissimo range, advice on each lesson, and a bibliography on the subject; Saxo Overtones E♭: app for playing harmonics to help the saxophonist gain a more solid foundation in his or her normal tone and acquire the required control and flexibility to perform the altissimo register; allows the user to record exercises; Saxo Overtones B♭: app for playing harmonics to help the saxophonist gain a more solid foundation in his or her normal tone and acquire the required control and flexibility to perform the altissimo register; allows the user to record exercises; Saxo Guide: method for working the saxophone including long tones, overtones, chromaticism, low register, middle register, and upper register; allows the user to record exercises; contains a tuning fork (A3, 440), advice on each lesson, and a bibliography on the saxophone; (iOS, varies by app).

Saxophone Fingering Chart (http://www.critical-apps.co.uk) Reference application that shows all the common fingering and trill combinations available on the saxophone including more than five hundred different combinations for soprano, alto, tenor, baritone, and bass, from A3 to C8; scroll left or right along the staff to the note and tap to select; tap the note again or use the back button to return to the staff; for fingerings with additional information, tap the

· 8 ·

String Instruments Apps

ALL STRING INSTRUMENTS APPS

Bowing Fun (http://www.bowingfun.com) Generates random string crossing exercises for violin, viola, cello, and double bass to develop sight-reading skills and agility; exercises are printable and can be individually programmed for any level of difficulty; included is a custom option for use in developing finger strength and picking speed on the guitar and other string instruments; (iOS 4.2 or later).

Exam Scales for String Instruments (http://www.mezzo-music.com/#!abrsm-strings-exam-scales/c1t44) Exam scales for violin, viola, and cello based on the ABRSM exam syllabus for grades one through eight; scales are categorized into grades, types, and octaves for easy navigation; high sound quality; training perfect pitch; intonation accuracy; (iOS 5.1 or later).

Fingering Strings (http://patrickqkelly.com/index.php/ipad/fingering-charts/fingering-strings) Fingering charts for violin, viola, cello, and double bass; pick a written note by touching the staff; finger placements are displayed and the concert pitch is played on the piano; select to display each of the fifteen positions for the violin and viola, sixteen positions for the cello, and either twelve Simandl positions or six Rabbath positions on the double bass; displays the placement of beginner tape on the fingerboard for all the instruments as well as string names and colors; view notation in treble, alto, tenor, or bass clef; always plays the concert pitch for any written pitch; double bass sounds an octave lower than written; sounds are included from C0 (below) to C9 (above) outside the range of the piano; touch and drag up and down on the staff to select the note; slide right for sharp, left for flat, or slide up and down near either edge for constant sharps or flats; when playing the keyboard, swipe to move the keyboard, tap to play notes, touch and hold, then slide to glissando; glissando up, notes will be notated in sharps; glissando down, notes will be notated in flats; (iOS 7.0 or later).

Get-Tuned Instrument Tuner (http://www.get-tuned.com) Supports and is able to tune the following instruments: banjo, bass guitar, cello, guitar, mandolin, ukulele, viola, and violin; supports alternate tunings and the ability to create and edit custom alternate tunings; option to display notes as sharps or flats; option to play tones once or repeatedly; can cycle through the strings automatically; two tones to choose from: the natural tone the instrument creates or a computer-generated tone; (iOS 4.3 or later; Android 2.2 and up).

RealOrchestra (http://www.mezzo-music.com/#!realorchestra-app/c10fk) String instrument apps for violin, viola, and cello; recorded in a real orchestra setting; (iOS 5.0 or later).

String Tuner (http://eumlab.com/string-tuner/) Chromatic tuner designed and optimized for string instruments: violin, viola, cello, and double bass; reference sound/tone generator is provided to help user tune by ear; tuning range from A0 to C8; (iOS 5.0 or later).

BANJO APPS

Banjo Companion (http://www.infinautgames.com/apps-for-musicians/banjo-companion/) Includes a metronome, drum tracks, custom themes, an ad-free option, and landscape mode in iPad; tool for any banjo player; banjo tuner; large list of chord charts and scales; beats to practice; simple interface; Tuner helps the user learn to tune by ear by clicking a peg to start or stop; Chord Charts show multiple positions for each chord including more than 1,900 chords; Scales includes more than 150 scales from blues and basics to more obscure charts; Alternate Tunings shows a wide variety of alternate tunings; with Show all Chord Notes, the user can view all the possible notes at once and build chords; Display Options show finger positions, intervals, or notes; with Auto Strum the user can hear the selected chord; with Manual Strum the user can play a chord by swiping across the strings; Dynamic String Indicators show which string or note is playing; Metronome shows basic beats with time signature options; Drum Packs; (iOS 3.0 or later).

Banjo Tuner Simple (http://www.tuneinstrument.com/products/tuners/banjo-tuner-simple.html) Banjo tuner used for the standard five-string Banjo in GDGBD tuning; (iOS 4.3 or later).

Banjo!!! (http://madcalfapps.blogspot.com) App includes the five-stringed banjo and six-stringed banjo instruments; Strum mode to play the banjo by strumming or tapping; Tap mode to play by tapping and sliding up and down the fretboard; Low Latency OpenAL Sound Engine; floating chord selector; bass guitar on/off option; (iOS 5.0 or later).

CELLO APPS

Cello Pro (T3 Apps) Graphics and sound quality for a unique cello experience; touch-screen capabilities; for all ages and levels; customize cello to individual preferences; built-in metronome adds beats in the background to aid in timing; compose an original cello piece; (iOS 4.0 or later).

Cello Tuner (http://jameswragg.com/cellotuner/) App to help hand tune cello; gives real pitch-perfect recorded acoustic cello sounds; can tune to the standard CGDA tuning or switch between four popular alternate tunings; listen to a single note or loop the playback; detailed tips on tuning and the anatomy of the cello; for beginners and experienced cellists; (iOS 3.1 or later).

Cello Tuner Simple (http://www.tuneinstrument.com/products/tuners/cello-tuner-simple.html) Cello tuner for the standard CGDA cello tuning; touch the note for the string to tune and turn the tuning pegs on the cello to match the note that is being played; can use for the viola as it is the same tuning, but the strings on the viola are one octave higher; make several passes tuning each of the strings; (iOS 4.3 or later).

Easy Cello—Cello Tuner (Games and Study Tools) Cello tuner app; tune cello quickly and efficiently for orchestra recitals or lesson practice; uses a combination of probability theory and Fourier transforms to provide the best chromatic tuner for string, woodwind, or brass instruments; aimed at those who play the cello; sample notes for C String, G String, D String, and A String, all at the right octave; (Android 2.2 and up).

PlayAlong Cello (http://www.atplaymusic.com/our-apps/playalong-cello/) Listens to playing and guides user through the selected song; choose the settings that complement skills; cumulative statistics of the total correct notes, consecutive correct notes, and the songs played correctly are available; beginners can choose to have finger charts displayed under each note on the staff as they play; touching the finger chart lets the user hear that note; pushing the play button allows the user to hear the whole song; advanced users can choose to display only the note names under the staff or to see a full screen of music; in Manual mode the user can play the song with their own interpretation, using the display like traditional sheet music; Stats display presents a list of all the songs, showing which have been played and how many of the notes the user was able to play; History feature keeps track of scores for past performances for each song played; Transposition feature lets the user shift the pitch of a song's notes up or down to customize the range of the song; in Game Center stats are listed on the leader boards and songs are presented as accomplishments; includes free songs and a chromatic scale; song list includes ratings of easy, intermediate, and advanced; share by e-mailing performances; (iOS 5.1 or later).

DOUBLE BASS APPS

Double Bass Tuner Simple (http://www.tuneinstrument.com/products/tuners/double-bass-tuner-simple.html) Tuner for the standard EADG double bass tuning; click on the note for the string to tune and turn the tuning pegs on the double bass to match the note that is being played; keep the tension on each string equal; bridge is not glued or attached to the double bass and is held there by the tension of the strings; if the tension varies too much, it might cause the bridge to collapse; while tuning, the bridge should always be perpendicular to the double bass and be straight, not angled; make several passes tuning each of the strings; (iOS 2.2.1 or later).

iDouble Bass (http://www.idoublebass.com/iDouble_Bass/iDouble_Bass_Home.html) App has dedicated section groups for beginners, bowing techniques, rockabilly style, swing style, and general lessons; includes 224 video lessons; Guided Tour to see how to play the lessons, alter the lesson title, alter the lesson description, add own user notes, give own rating to each lesson, skip backward and forward between the lesson groups; (iOS 3.2 or later).

Learn Double Bass (Rico7) App includes video lessons on how to play the double bass; tips and tricks; (Android 2.1 and up).

Learn to Play the Double Bass (Easysource HK) App includes 114 video lessons on how to play the double bass; (Android 2.3.3 and up).

My Double Bass (Musical Mania) Composing app; includes double bass notes, double bass tuner, and double bass chords; can select background music and sounds to accompany playing; high-quality sound; different variations of sound editing; (Android 2.2 and up).

Virtual Double Bass (Yuki Yazilim) Virtual double bass application; (Android 2.2 and up).

DULCIMER APPS

Dulcimer (http://www.gclue.com/?aid=016) Record and play functions; store recording data into PC in MIDI and WAV format; practice mode; demonstration tunes; (iOS 5.0 or later).

Dulcimer Tuner Simple (http://www.tuneinstrument.com/products/tuners/dulcimer-tuner-simple.html) Dulcimer tuner for the standard DAA dulcimer tuning; (iOS 3.0 or later).

Hammer Dulcimer (http://www.savageapps.com) Hear sounds that echo from deep within the middle ages, where the dulcimer was born; believed to have been invented in ancient Persia; the name dulcimer comes from the Latin word *dulcis* (sweet) and the Greek word *melos* (song); used in Celtic and folk music of current times; easy to play; every note fits into the key and scale selected; (iOS 4.1 or later).

HARP APPS

Air Harp (touchGrove, LLC) App reproduces the sound of a harp; virtual harp has fifteen strings representing two octaves of notes in G Major; can alter each string and play it sharp or flat by tapping and holding the corresponding key on the left panel; tap the strings to create own tune; use sheet music to play a well-known song; includes several pieces of sheet music; uses real notes and strings; multi-page music support for sheet music; pick a song, slide the music under the strings, and pluck the notes along the strings according to the diagram; includes sheet music for ten popular songs with lyrics; (iOS 7.0 or later).

Harp (http://harpapp.com/) Software instrument; left hand picks chords and right hand strums notes in a four-octave range; can play 168 different chords including fourteen chord structures in any of the twelve keys; sing along with the app in live performance; make song transcriptions; educational tool for studying intervals, chords, and chord progressions; (iOS 3.0 or later).

MANDOLIN APPS

Mandolin (http://www.yonac.com/software/mandolin/index.html) App has 44100 Hz, sixteen-bit sound quality; zero latency; three different instruments including acoustic mandolin 1, acoustic mandolin 2, and electric mandolin; tapping and plucking algorithms; three picking modes include Tap & Drag or Classic Mode, Pluck Only or Trem Picking Mode, and Tap & Pluck; adjustable plucking sensitivity; reverb unit with adjustable time and feedback; three reverb presets; configurable orientation; right-handed or left-handed layout; adjustable fret width, pick-area width, and string spacing; each of the eight strings are individually tunable; tune while testing; ten common and uncommon mandolin tunings built-in; capo option for the lowest visible fret; fretboard scroll feature; moveable menu; (iOS 3.0 or later).

Mandolin!!! (http://madcalfapps.blogspot.com) App includes two mandolins, one standard and one baritone; realistic sound samples and response; two modes for jamming including strum and tap; bass guitar blend option; chords, keys, chord voicings; right- and left-hand layouts; Low Latency OpenAL Sound Engine; floating chord selector; (iOS 5.0 or later).

Mandolin Tuner Simple (http://www.tuneinstrument.com/products/tuners/mandolin-tuner-simple.html) Tuner for a standard mandolin in GDAE tuning; (iOS 4.3 or later).

UKULELE APPS

My Ukulele (http://trajkovski.net/sound.html) Virtual ukulele simulation for Android; play uke on device; two ukulele types include standard uke and banjo uke; four play modes include solo mode, tapping mode, chord mode, and combined mode; includes all 228 chords; studio-quality sound; multi-touch support; six sound effects; play velocity; string muting; string damping; string bending; string animation; right- and left-handed; seven neck skins; four, five, or six frets; a capo option; virtual vibrato arm; virtual volume knob; chords load/save; SD card support; tablet/pad support; hardware acceleration; advanced haptic feedback is provided by Immersion's MOTIV engine; (Android 2.1 and up).

Play Ukulele (http://ukulele.u2app.net) Turn smartphone into a ukulele; practice and try various chords; video guidance; chord finder/chords display; editing/setting chord; real ukulele tunes; playing and switching saved chords in one screen; tuning mode; (iOS 5.0 or later).

UChord (http://www.asoft.ne.jp/iOS/en/) Chord book for ukulele; (iOS 5.1 or later).

Ukulele Companion (http://www.infinautgames.com/apps-for-musicians/ukulele-companion/) Tool for any ukulele player; ukulele tuner; beats to practice with on iPhone, iPad, or iPod; simple interface; tuner for tuning by ear; chord charts including more than 1,900 multiple positions for each chord; more than 150 scales from blues and basics to more obscure charts; display options include finger positions, intervals, or notes; with auto strum user can hear the selected chord; with manual strum user can play the chord by swiping across the strings; dynamic string indicators show which string or note is playing; metronome shows basic beats with time signature options; variety of drum tracks available; (iOS 3.0 or later).

Ukulele Practice (http://www.iphone-ipad-iapp.com/iphone.html) App includes the 144 main chords of the ukulele, a metronome, and a tuner; Universal app; (iOS 7.0 or later).

Ukulele Toolkit (http://eumlab.com/ukulele/) Collection of tools for ukulele players including: pro-level tuner; metronome; strum patterns to learn how to strum; 116 drum loops to jam along; scales with 108 different styles; interactive chord diagrams showing one chord in different frets; circle of fifths; transposition function; backward chord search tells user the name of a chord spontaneously discovered; learn and play along with selected chord progressions; left-handed supported; capo helps user search and learn chords under capos; changeable A4; supports tenor, soprano, and baritone tuning modes; (iOS 6.1 or later).

VIOLA APPS

Easy Viola (Games and Study Tools) Viola tuner app; use for orchestra recitals, a music lesson, or practice; uses a combination of probability theory and Fourier transforms to provide the best chromatic tuner for string, woodwind, or brass instruments; aimed at those who play the viola; includes sample notes for C String, G String, D String, and A String at the right octave to ensure speed and accuracy; guiding notes; auto-tuner; (Android 2.2 and up).

PlayAlong Viola (http://www.atplaymusic.com/our-apps/playalong-viola/) Listens to playing and guides user through the selected song; choose the settings that complement skills; cumulative statistics of the total correct notes, consecutive correct notes, and the songs played correctly are available; beginners can choose to have finger charts displayed under each note on the staff as they play; touching the finger chart lets the user hear that note; pushing the play button allows the user to hear the whole song; advanced users can choose to display only the note names under the staff or to see a full screen of music; in Manual mode the user can play the song with their own interpretation, using the display like traditional sheet music; Stats display presents a list of all the songs, showing which have been played and how many of the notes the user was able to play; History feature keeps track of scores for past performances for each song played; Transposition feature lets the user shift the pitch of a song's notes up or down to customize the range of the song; in Game Center stats are listed on the leader boards and songs are presented as accomplishments; includes some free songs and a chromatic scale; song list ratings of easy, intermediate, and advanced; share by e-mailing performances; (iOS 5.1 or later).

Viola Tuner (Alvin Yu) Plays tuning notes for viola (C, G, D, A); (iOS 7.0 or later).

Viola Tuner (http://www.123violatuner.com) Real strings from a viola to help tune by ear; (Android 2.2 and up).

VIOLIN APPS

Classical Violinist (http://apps.h2indie.com) Improve musical and coordination skills with a challenging violin game; gives hands and fingers a workout; play along with prerecorded classical violin music; earn points by matching the signals with bowing actions; glide the bow back and forth on the virtual strings like using a real violin bow; included are fifteen beautiful violin pieces; more songs can be unlocked within the app; (iOS 4.3 or later).

Fiddle Companion (http://www.infinautgames.com/apps-for-musicians/fiddle-companion/) Tool for any fiddle or violin player; intuitive interface; tuner helps user learn to tune by ear by clicking a peg to start or stop; chord charts show multiple positions for more than 1,600 chords; includes more than 150 scales from blues and basics to more obscure charts; display options include finger positions, intervals, or notes; with auto strum user can hear the selected chord; with manual strum user can play a chord by swiping across the strings; string indicators show which string or note is playing; metronome shows basic beats with time signature options; variety of drum tracks available; landscape mode; localized for sixteen languages; (iOS 3.0 or later).

Fiddlicator (http://fiddlicator.wordpress.com) Audio tool that can simulate various kinds of acoustic environments by the convolution of the input signal with a custom impulse response; was developed to simulate an acoustic body resonance for electric musical instruments, mainly the electric violin; can also be used for a cab simulation with proper impulse response files; files can be imported via iTunes share or via "Open In" option from another application; demo videos; main filter is based on convolution with filter kernel up to 16k TAPs; own custom IR kernel files can be imported; six-band parametric equalizer including low-shelf, 4xpeak, high-shelf; simple delay; simple reverb; high-pass and low-pass filter on input; low latency; mono input/output 44.1 kHz allows channel selection; Audiobus and IAA support; (iOS 5.1.1 or later).

Irish Fiddler (https://www.facebook.com/irishfiddlerapp) Mairéad Hickey coaches students as they learn to play Irish tunes on the fiddle; for anybody with an interest in Irish fiddle playing, violin, Céilí, or Celtic music; each tune contains a listening lesson, tutorial, slow playthrough and bowing instruction, and printed music; includes a tutorial of how to transition from one tune to another so that the tunes can be played as a set or medley; (iOS 3.0 or later).

Learn to Play Violin (KRM Apps) Includes how to purchase a violin; basic violin techniques; how to tune the violin; interact with other violin players and get help; basic video instructions; a written violin guide; includes videos and written material; (Android 2.2 and up).

Learn Violin (http://www.mahalo.com) Includes helpful exercises and simple techniques to remember violin fingering; learn how to string and maintain the violin; keep track of the instrument's many parts; correct bad habits; learn music theory including major chords and scales; how to get the best tone; taught by Paul Dateh; breaks down violin techniques, including tuning, basic notes, beginning songs, and more; play along with him to learn the important skills all beginners should know; (iOS 5.0 or later).

Learn Violin (Rico7) Learn the violin; violin tips on video; violin tutorial videos; children's violin lessons; violin lessons for beginners; (Android 2.2 and up).

nTune: Violin (http://www.cocoada.com/apps.html) Written by a violinist that needed an accurate application for tuning; uses actual recorded violin notes; playback options include arco (bowing) or pizzicato (plucking); contains the basic tuning of G, D, A, and E; push the button for the corresponding note; includes settings for playing one note with each click or a continuous loop with silence between to allow user to tune; (iOS 5.1 or later).

PlayAlong Violin (http://www.atplaymusic.com/our-apps/playalong-violin/) Listens to playing and guides user through the selected song; choose the settings that complement skills; cumulative statistics of the total correct notes, consecutive correct notes, and the songs played correctly are available; beginners can choose to have finger charts displayed under each note on the staff as they play; touching the finger chart lets the user hear that note; pushing the play button allows the user to hear the whole song; advanced users can choose to display only the note names under the staff or to see a full screen of music; in Manual mode the user can play the song with their own interpretation, using the display like traditional sheet music; Stats display presents a list of all the songs, showing which have been played and how many of the notes the user was able to play; History feature keeps track of scores for past performances for each song played; Transposition feature lets the user shift the pitch of a song's notes up or down to customize the range of the song; in Game Center stats are listed on the leader boards and songs are presented as accomplishments; includes some free songs and a chromatic scale; song list includes ratings of easy, intermediate, and advanced; share performances; (iOS 5.1 or later).

Real Violin (http://phyar.cn) Simulates violin playing; real-time multi-track sound processing engine, high quality, low latency; Slide mode and Touch mode; with Slide mode, user can play the "Real Violin" like a real violin; with Touch mode, user can play the sound directly by tapping the corresponding note, like playing a keyboard; slide along the string with the finger of the left hand to play an unlimited smooth glide; custom distance between the note marks; includes three violin family instruments (violin, viola, cello) in one; up to one octave per string with Touch mode; easy to use interface; (iOS 4.2 or later).

The Violin Gallery (http://www.weinreichlabs.com) Informative and entertaining app lets user explore biographies, sound samples, detailed photographs, and interactive images of many fine violins; for musicians, teachers, artists, history buffs, and art lovers; view dozens of images of three-hundred-year-old instruments made by the world's greatest craftsmen; swipe to see violins in full-screen 360-degree views; tilt the iPad to see how light changes the shape and color of the wood's grain; read histories for each instrument; biographies of all the makers; recorded audio samples from each instrument; (iOS 6.1 or later).

Tiny Violin (http://www.gorbster.net/blog/tiny-violin) Tiny violin for iPhone or iPod Touch; select one of three violin sound types; "Despair" displays sincere condolences with a weeping melody; "Joy" celebrates the good times with a light jig; "Let's Rock!" plays an electric riff; more than fifteen total sounds; play by sliding bow with finger or by shaking device; sound samples provided by efiddler.com; (iOS 6.0 or later).

Vamoosh Violin (www.vamooshmusic.com) *Violin Tutor Book 1*; fun introduction to violin playing; includes backing tracks; simple format; example videos; compatible books for viola, cello, and double bass are also available; (iOS 4.0 or later).

Violin (http://www.egert.us/#violin) Violin is the most realistic violin, viola, cello, and double bass app for Android devices; play by bowing (arco) or plucking (pizzicato); adjustable layout fits both phones and tablets; (Android 2.1 and up).

Violin!!! (http://madcalfapps.blogspot.com) App for the iPad; has a violin, cello, and viola for playing in three modes; in Slide Mode, user plays the full range of the violin by sliding or tapping fingers up and down the neck as the other hand plays the strings; in Orchestra Mode, the violin, cello, and viola are combined for full chord playing; in Tap Mode, screen is filled with buttons to tap and play the full range of the instruments; iPad tilt vibrato; attack and release controls; realistic-sounding instruments; (iOS 5.0 or later).

Violin Lesson Tutor (http://www.amsmusictuition.co.uk) App takes user directly to YouTube Channel; covers the basics of learning to play the violin with the addition of grades one through five theory; lessons are delivered by professional musician and teacher Alison M Sparrow; for beginners through anyone who wants to refresh their knowledge; (Android 1.6 and up).

Violin Lessons (http://florencealtenburger.com) Designed for teachers, parents, and students; provides a series of exercises that the student and teacher can work on during the lesson; after the lesson, the pupil and parent can use the app to continue to work on all aspects of the exercises; not designed to replace a violin lesson but to enhance it and provide the teacher with material for the pupil to work on at home; (iOS 6.0 or later).

Violin Multi-Tuner (http://martinbritz.blogspot.com/2011/04/violin-multi-tuner.html) App for violinists; fast and accurate pitch detection; tone generator with twelve presets; accurate metronome; playable fingerboard; bowed and pizzicato sound; ear and sight-reading exercises; eight scales and seven modes in seventeen keys; 3-D accelerated graphics; pitch detection graph; Universal app; (iOS 6.0 or later).

Violin Tune (http://www.code-app.com/apps) Simple, streamlined, stylistic, efficient, highly accurate violin tuner; hears the sound of the violin and tells user how to adjust the string to match the exact sound; open app, play a string, adjust, done; supports standard violin tuning (G-D-A-E); red wood theme to match the violin; (iOS 3.1 or later).

Violin Tuner (Alvin Yu) Plays tuning notes for the violin (GDAE); (iOS 7.0 or later).

Violin Tuner Simple (http://www.tuneinstrument.com/products/tuners/ violin-tuner-simple.html) App is a simple violin tuner for a standard violin in GDAE tuning; (iOS 4.3 or later).

Virtual Violin (https://sites.google.com/site/snowysapps/) Simple user interface; includes thirty-two notes; three play modes; different backgrounds; (Android 2.2 and up).

· 9 ·

Drums and Percussion Instruments Apps

DRUM KIT APPS

Classic Drum (http://www.rodrigokolb.com.br/#apps) Real drum kit with acoustic percussion sounds; play live music; multi-touch; thirteen drum pads; complete acoustic drum kit; fifteen realistic drum sounds; studio audio quality; includes kick, bass, snare, tom, floor, cymbal, hi-hat, ride, crash, bell, cowbell, and tambourine; ten examples of rhythms; record mode; play in loop; rename recordings; works with all screen resolutions; for drummers, percussionists, musician, performers, and artists; (Android 2.1 and up).

Drum Kit Master (CouldSys) Application for playing the drums; multi-touch screen; contains several kinds of different drum sounds; supports all screen resolutions; record own music using the menu on smartphone; listen to recordings; play new songs; (Android 1.5 and up).

Drum Kit Plus (CouldSys) Multi-touch drumming application; experience finger drumming; contains thirteen different types of drum sounds; supported by all screen resolutions; record and listen to recordings and new songs; (Android 1.6 and up).

Drum Kit Pro (http://nullapp.com/details.php?id=1) Advanced drum app; features more than twenty different drums, cymbals, and percussion elements; move drums across the screen, adjust each drum volume and screen size separately, and add or remove drums from setups; create and edit own drum setups and record music beats; share with friends; (Android 2.3 and up).

Drum Kit Pro (http://www.getdrumkit.com) Make beats; play along to songs in iPod library and record own versions; layer tracks on top of each other; add beats, cymbals, and fills to existing songs; create original songs; fast response on all devices; no delay or latency when playing drums; updated frequently with new sounds, themes, and features; six-piece kit including four toms, snare, bass, hi-hat open and closed, crash, ride, splash; acoustic, electronic, and industrial themes; five different kits include classic, rock, hip hop, techno, and dance; visual feedback when drum head is tapped; professionally recorded, high-quality drum sounds; play multiple drum heads simultaneously; slide finger back and forth between bass drum and pedal for fast double bass for metal songs; record, load, and edit tracks; with metronome can tap bpm to set or set manually; to record, tap the info button on the bottom right; tap the record button and play a beat; tap the stop button when done and give track a name; can play on top of a track while it's playing for multi-track recording; metronome and count-in for easier recording; can slide fingers across to play rolls; turn on rolls in Settings and change the bpm (beats per minute) to fine-tune rolls; Jam Packs; Electronic Theme Pack with two drum pad-style themes including a retro electronic theme and a harder-edge industrial theme; Kit Remains The Same is a 1970s-style vintage kit; includes new kit graphics and studio recorded sounds; includes drum animations that pop out and a kick pedal that moves; (iOS 6.1 or later).

Drum Meister (http://www.greysox.net/app-list/drum-meister) Realistic drum-playing environment; four different drum kit sounds including rock, jazz, dance, and electronic; ten basic drum components with twelve different sounds played simultaneously; eight extra components include four brushed, side-stick, two extra cymbals, and a cowbell; arrange drum components to make fingers comfortable; edit components; auto drum rolling (four, eight, twelve, sixteen, thirty-two beat); mixes high quality polyphony sounds with no clipping at the end; 3-D positioning sounds; metronome with adjustable on/off, bpms, and beats; open hi-hat features quick stop; record and save own beats; forty pre-recorded beats include rock, hip hop, jazz, Latin, ballad, dance, etc.; slide and play; activates motion sensor for playing bass drum and cymbals; responsiveness for double bass and fast beats; animated visual effect; velocity control; (iOS 3.0 or later).

Drum School (http://drumschoolapp.com) Advanced groove library and drum learning tool for all levels; tool for advancing and expanding skills; more than 260 drum grooves in many different styles; break down groove into its basic components; includes twelve video demonstrations of basic drum technique for hands and feet; video performances by world-renowned drummer Ferenc Nemeth; practice section includes 133 drum exercises; standard drum notation with high-quality audio; collect favorite grooves and ones currently working on; difficulty level from one/easy to six/hard; track practice progress; (iOS 7.0 or later).

Drum Set (http://www.nullapp.com/details.php?id=2) Record own beats; features different drum set variations with drums and percussions; supports multi-touch; animations; five different drum layouts; each layout consists of a different set of drums and cymbals; wide range of splashes, rides, crashes, hi-hats, snares, toms, and more; layouts currently include: basic drum set, big concert set, jazz set, double bass set, and electric drum set; components include kick, bass, snare, tom, floor, cymbal, crash, splash, ride, wood, cowbell, electric drum pad, and different percussions; each set has its own sounds, look, and feel; (Android 2.3 and up).

Drum Solo HD (http://www.batalsoft.com/#!apps/cngp) Multi-touch acoustic drum kit simulator; play with fingers in mobile phone or tablet; fast response; includes different sound packs recorded with studio quality;

record songs and share with friends; play with a headset for a better experience; designed for everyone including children, percussionists, musicians, and drummers; eleven demo songs to learn to play drums; drag finger for different drums and play a solo; video sample; choose between four exclusive sound packs: classic rock, heavy metal, jazz, and synthesizer; touch up to twenty fingers simultaneously; reverb effect simulates a live performance; record own session and play on it like a real drum machine; record, play, and repeat compositions; record un unlimited number of notes in loops; realistic sampled stereo sound, including double kick bass, two toms, floor, snare, hi-hat in two positions with the pedal, splash, crash, and cymbal; HD drum images; double bass drum pedal available; animations for each instrument; repeat button to play continuously with improvisations; low latency; eleven touch-sensitive pads; fast loading time; (Android 2.2 and up).

DrumPerfect (http://www.drumperfect.nl) Drum sequencer; creates natural-sounding drum tracks; three extensive editors to create human-like drum sequences; (iOS 7.0 or later).

Drums & Percussion (http://www.drumsundpercussion.de//) Includes entertaining and informative in-depth interviews with the world's leading protagonists of the drummer's scene and exclusive photographs from stages and clubs; in "Drumming Pioneers" focuses on musicians that laid down the tracks for today's drumming scenery; covers events, manufacturers, reading, and photos; reviews of the latest CDs, DVDs, and books as well as stories about selected subjects; workshops and practical advice sections from leading drummers; display of pure text in variable sizes; navigation via graphical table of content or a menu with list view; multimedia and online content can be accessed by a fingertip through self-designated areas in the layout; can scroll, zoom, and navigate in the magazine with iPad on the touch screen; more than the electronic version of the magazine; in addition to content of the print edition there are many videos, slide shows, and audio samples; free multimedia experience for every drummer; (iOS 3.2 or later).

Drums! (http://www.cinnamonjelly.com/drums.html) Realistic drum sounds; record and export sessions to other popular apps like GarageBand; includes three drum kits that come as standard with two other kits available via in-app purchase; play along to songs in music library using the built-in music player; record drumming alone or along to music; choose from music library or a previously recorded track; export recordings via e-mail, WiFi, Dropbox, SoundCloud, or AudioCopy; share recordings between devices; left- and right-handed drum layouts; play with a single- or double-kick drum setup; switch out the second kick drum for either a cowbell, tambourine, or wood block; dual zone ride cymbal with regular or crash sounds; hit the snare drum rim to trigger a rimshot; low latency; option to trigger the kick drum when user hits the crash cymbal; high-quality Retina graphics; Universal app; for all levels; (iOS 7.0 or later).

(Drums) Percussion Instrument (CouldSys) Set of drums; multi-touch screen; create own songs; supported on all screen resolutions; (Android 1.6 and up).

Epic Drum Set (http://mobafun.com) Feel the beat of drums; record own beats and play them back; save recordings; choose any song in iTunes library and play along; create, record, and play along with own beats; choose from a range of four different drum sets including two free, standard, industrial, hard rock, and junk; each sound is professionally recorded and provides a realistic sound; (iOS 4.3 or later).

Kid's Finger Drums (http://www.goatella.com/apps/music/finger-drums) Lets kids play around with drums without taking up the space or making the noise of a drum set; encourages creativity and musical interest; (iOS 4.3 or later; Android 2.0 and up).

Pocket Drums and **Pocket Drums 2** (http://www.codyrotwein.com/index.html) Includes high-fidelity stereo drum kits; three different drum wraps; realistic drum graphics and visual effects; unique drum loops to play along to; variable pitch control and delay; drum rolls; variable metronome with ten unique sounds; lefty mode; multi-touch to play multiple drum heads and cymbals simultaneously; multi-track recorder; save, load, loop, and play recorded tracks; record and play on top of loaded/saved tracks; pitch and delay controls; slider adjustments allow limitless drum sounds; create realistic rolls by swirling finger around to create slower or faster rolls; double bass drum feature with slider adjustment; ten metronome sounds and four time signatures to choose from with bpm adjustment; ten drum beats and ten music beats to play along to; option to add bass drum when hitting crash cymbal; videos of basic drum rudiments played on a real snare drum; repeated cymbal and drum taps blend in like real drums; play along to music in library; realistic drum wraps and visual effects; drums and cymbals look and respond like a real drum set; (iOS 4.3 or later; Android 2.2 and up).

Real Drum! (http://www.rodrigokolb.com.br/#apps) Complete drum kit with acoustic percussion sounds; multi-touch; thirteen drum pads; nineteen realistic drum sounds including kick, bass, snare, tom, floor, cymbal, hi-hat, ride, crash, splash, bell, china, block, cowbell, and tambourine; studio audio quality; ten examples of rhythms; record mode; play in loop; rename recordings; works with all screen resolutions; for drummers, percussionists, musicians, performers, and artists; (iOS 6.0 or later; Android 2.1 and up).

Real Drums (http://gismart.com) App makes a real drum kit out of device; realistic graphics; for amateurs as well as experienced drummers; includes many popular drum kits (rock, hip-hop, jazz, dance, and more); metronome; special sound settings for each instrument; different visual styles; fast response time without delay; HiFi sounds; animated visual effects; background music; (iOS 5.0 or later; Android 2.2 and up).

Spotlight Drums (http://www.pocketglow.com) Formally known as 3D Drum Kit; animated 3-D drum kit on a real stage; includes ten drum kits, each with a bass drum, snare drum, three tom drums, a hi-hat, two crash cymbals, a ride cymbal, and a cowbell; color-changing stage lighting and fog effects; drag, zoom, and position drum from any angle; perform drum rolls with a swipe; optimized for all iOS devices; record and save performances; export recordings as MIDI files via iTunes File Sharing; watch drum kit play itself as it plays back recordings; access to fifty built-in, loopable demo beats in a variety of different genres; adjust the playback tempo of recordings and the built-in demos; (iOS 5.0 or later).

WORLD MUSIC PERCUSSION APPS

Bongos—Dynamic Bongo Drums (http://www.skunkbrothers.de/lang/en/bongos-dynamic-bongo-drums/) Professionally recorded and mastered sounds; bongos' sound varies depending on where hit; eighteen high-quality bongo samples; no latency for fast and exact playing; optimized for iPhone and iPad speakers, Android speakers, headphones, and external speakers; supports multitasking; play along with own music; (iOS 2.5.1 or later; Android 2.1 - 2.3.7).

Brazil Samba Percussion (CouldSys) Composed of fifteen instruments to play with fingers on smartphone; includes the rhythms of percussion sounds from Brazil; multi-touch in some versions of Android; can record the sounds; for all ages; (Android 1.5 and up).

Cody Rotwein: Drum Apps (http://www.codyrotwein.com/index.html) In addition to Pocket Drums, the drum apps available include: Bongos!,

Bongos! Bongos!, Cajon!, Congas!, Conga Drums, Djembe!, Doumbek!, Drum Kidz, Frame Drum!, Hand Drums, iNatureSounds, Play Drums!, Rockin' Drums, Shamanic Journey, and Tabla!; (iOS, varies by app).

Congas and Bongos (http://www.rodrigokolb.com.br/#apps) Drum pad with Latin percussion sounds; play congas or bongos; play live music; multitouch; fifteen realistic drum sounds; studio audio quality; real percussion set; instruments like conga, bongo, crash, cymbal, chimes, and vibraslap; record mode; play in loop; rename recordings; works with all screen resolutions; for drummers, percussionists, musicians, performers, and artists; (iOS 6.0 or later; Android 2.1 and up).

Djembe Drum (DAPPS) Feel the rhythm of the African drum djembe on the iPhone; two different sound variations include normal and electronic; (iOS 1.1 or later).

Drums by Asrodot (http://www.asrodot.com/drums/) App comes loaded with six drum kits tailored for various styles of music including cajón, xylophone, jazz, IDM, drum and bass; quick and responsive action; every pixel sounds a little different; velocity sensitive; can add dynamics by tapping harder or softer; increase gain to add a tube-like distortion that can dramatically change the sound; delay module, feedback, and mix settings; designed for both right-handed and left-handed drummers; with Audiobus support can stream live audio directly to other Audiobus compatible apps; (iOS 5.1 or later).

iTablaPro—Tabla Tanpura Player (http://upasani.org/home/iTablaPro.html) Electronic tabla and tanpura; for Indian classical musicians and students; includes support for all common tablas used in Hindustani music; companion for daily riyaz practice; tanpura and swar mandal create a concert atmosphere; Audiobus supported input; (iOS 5.0 or later).

Latin Percussion (http://www.iphone-ipad-iapp.com/ipad.html) App that allows user to play Latin and Cuban percussion instruments; includes bongos, congas, timbales, cowbells, agogos, tambourine, chimes, and cymbals alone or with music content; (iOS 4.2 or later).

Monkey Drum (http://flippfly.com/monkeydrum/) Jam on a realistic drum and watch as jungle animals copy the rhythm; watch funny reactions while feeding them bananas, spinning them, or bopping them on the head; make them happy and they will reward with a special song; create music by drawing notes on a colorful grid and watch as the creatures play it; tap a button to make the song into a music video or remix the song with different instruments and tempos; save and name songs; 3-D graphics; djembe drum, marimba, and kalimba included; earn heart coins and bananas by playing; unlock new instruments and other fun items including microphone, congas, girl monkey, panda, and bunches of bananas; (iOS 4.3 or later).

Percussion Pack (http://www.YouTube.com/watch?v=1bR5jsDoOWI &feature=c4-overview&list=UU7MYUjC28QCSK5VwWL4IcEw)   App gives users eight exclusive world music percussion instruments with high-definition live sound quality; each instrument was professionally recorded to give users the ability to play from their mobile device and re-create each instrument's sound in its true form; can use instruments and sounds for recording or live performance purposes professionally or for fun; (iOS 7.0 or later).

Percussive (Touch Media Productions) Contains a total of five instruments including glockenspiel, kalimba, xylophone, vibraphone, and marimba; 3-D designed artwork; learn, practice, and master music and touch skills; (iOS 3.2 or later).

Real Percussion! (http://www.rodrigokolb.com.br/#apps) Drum pad with Latin percussion sounds including congas, bongos, timbales, block, cowbell, and tambourine; play live music; multi-touch; twelve realistic drum sounds; studio audio quality; record mode; play in loop; rename recordings; for drummers, percussionists, musicians, performers, and artists; (iOS 6.0 or later; Android 2.1 and up).

Shakers (http://trajkovski.net/sound.html) Virtual shaker with ten different instruments; shake the device; press the pads to play or record; includes maracas, castanets, claves, sand eggs, woodblock, guiro, cowbell, and bell; studio-quality recorded sounds; shake force variable volume; device position variable tone; shake sensitivity adjustment; vibrate on shake simulates real shaker; continuous sound on hold option; user shaker sample recording; sample load and save; sample import from SD card; sample export to SD card; four global DSP sound effects; fifteen backgrounds; shake animations; tablet/pad support; (Android 2.1 and up).

· *10* ·

Piano and Keyboard Instruments Apps

PIANO APPS

Angel Piano (https://play.google.com/store/apps/details?id=com.Angel Piano&hl=en) Music game; play piano by following the instructions; for beginners; (Android 1.6 and up).

CMP Grand Piano (http://www.crudebyte.com/mobile/cmp_grand_piano/) High-quality virtual piano sound module designed for professional usage on stage by connecting an external MIDI keyboard to the device by using the Apple USB camera adapter or another adapter supported by Apple; large collection of audio samples; various articulations of each key of a real piano were sampled in full length and high quality; real-time disk streaming is utilized to playback the respective audio samples with very low latency; Jack Audio/MIDI Connection Kit supported; supports iOS Inter-App Audio; (iOS 4.3 or later).

Echo Piano Pro (http://supertintin.net/echopianopro.html) Recordable virtual piano; slide screen to reach all the keys during a performance; Apple Core MIDI compliant; (iOS 3.2 or later).

FingerPiano (http://fingerpiano.net/fingerpiano/) Allows users to play the piano with their fingers; instead of reading the score, scrolling guides appear on the screen; provides eighty-eight pieces of famous music; play with one hand or with both hands; (iOS 5.0 or later).

Go! Piano (http://bellstandard.com/?app=go-piano) Play the piano on mobile device; easy-to-use interface; record own songs and play them back; metronome features a wide range of bpm; select from six high-quality sound packs including Grand Piano, Music Box, "Ahh" Choir, Electric Piano, Accordion, and Banjo; (iOS 4.3 or later).

Grand Piano (http://www.scotchware.com/apps) Portable piano and on-the-go organ; versatile piano or organ app for iPad; choose to play one of seven instruments including organ 1, organ 2, music box, bell, string (guitar), piano, or electric piano; two octaves are displayed on the screen; option to switch to a higher octave with a total range of four octaves; customize the sound by turning the "echo" option on or off; (iOS 6.0 or later).

Grand Piano Virtuoso HD (http://amyfaulkner.yolasite.com) Sounds like a grand piano, organ, or synthesizer; echo effect to give a rich, lush, and bright tonality; tune to any octave to experiment and create; (iOS 4.3 or later).

I Love Piano (http://www.bungbungame.com/EN/apps/apps_page.aspx?appName=3) Realistic 3-D piano app; sound and keys modeled after a real piano; for anyone studying the piano or anyone wanting to start; 3-D keys and touch-screen ripple effect; tactile piano app; realistic keys; true-to-life sound; percussion and accompaniment; ascend/descend between eight registers; supports multi-touch; supports one or two keyboards; duet mode is supported for two keyboards; musical scale display; save/replay sessions; change screen settings; key width can be changed to suit preferences; (Android 4.0 and up).

iLectric Piano (http://www.ikmultimedia.com/products/ilectricipad/) Studio-quality electric piano app; collection of forty professional-sounding instruments for iPad including vintage electric piano and electronic keyboard instrument sounds; Audiobus support; (iOS 7.0 or later).

KeyPlay (http://icemediacreative.com) Piano app; multi-touch keys; optional key labels; two sounds; play along with music library; Apple Core MIDI compliant; (iOS 4.2 or later).

MiniPiano (http://fingerpiano.net/minipiano/) Piano app with fourteen notes; instead of reading the score, scrolling guides appear on the screen; (iOS 6.0 or later).

My Piano (http://trajkovski.net) Virtual piano for Android; studio-quality sound; eleven instruments; seven sound effects; multi-touch; note velocity; note aftertouch; note pitch bend; integrated sampler; integrated recorder; sample recording; sample load and save; track load and save; MIDI load and save; MIDI over WiFi; MIDI note velocity; MIDI note aftertouch; MIDI volume control; MIDI pitch bend; one and a half or two octaves; tablet/pad support; hardware acceleration; looped playback; twelve note polyphony; sixteen piano skins; five key types; eight window themes; window animations; (Android 2.1 and up).

Pianist (http://moocowmusic.com/Pianist/index.html) First widely available mobile multi-touch piano; full eighty-eight key piano keyboard sampled from a real piano for authenticity; record and save compositions for later playback; overdub multiple times and remove unwanted notes; virtual soft and sustain pedals; expression depending on where the key is pressed; optional dual keyboard layout allows user to reach more keys without the need to scroll; up to four octaves can be displayed and played at once; configurable metronome; (iOS 5.0 or later).

Pianist Pro (http://moocowmusic.com/PianistPro/index.html) Photo-realistic graphics; full recording and overdub facility; live MIDI control of external synths via Core MIDI USB, MIDI Mobilizer interface, Network MIDI, OSC, or DSMI; standard MIDI file import and export; Audio (WAV) file export; sixteen professionally recorded instruments (pianos, organs, guitars, synths); built-in effects, arpeggiator, and drum machine; soft and sustain pedals; velocity based on key strike position; full eighty-eight-key single or dual keyboard layout with configurable key size; on-screen pitch bend, modulation, and swell pedal controls; vibrato and low-pass filter; full MIDI import and export capability, allowing user to record song ideas and export them via MIDI to a Digital Audio Workstation; can use as a master MIDI keyboard, either with USB, the Line 6 MIDI Mobilizer interface, or via WiFi; drive soft or hard synths wirelessly; can be used as a MIDI sound source, driven by an external hardware keyboard; built-in arpeggiator allows user to create perfectly timed runs by holding down a few keys; complex patterns of notes and chords can be programmed into the arpeggiator and played back live; play shuffle-beat blues piano with walking bassline and overlayed jazz chords by holding down a single chord; changing the chord changes the melody but maintains the tight rhythm provided by the arpeggiator; drum machine with a simple interface for adding rhythm; choice of drum kits; delay effect can be used to bring energy and interest to a synth bassline or a guitar or piano riff; "Scale Piano" interface allows user to select a musical scale from a large choice of presets (e.g., Natural Minor or Pentatonic Blues) or define own scales from scratch; notes and runs on a scale can be played back in any key by sliding a finger up and down a virtual touchpad on-screen; for musicians of all levels; website has demonstration videos, a full description of functionality including the user manual, more screenshots, and a support link; (iOS 4.2 or later).

Piano DX Free (http://www.betterdaywireless.com/BetterDayWireless/BDW.html) Eighty-eight-key piano; multi-touch; rich sound; adjustable metronome; sustain pedal; double row keyboard; Retina display graphics; choose piano key sizes to exactly fit fingers; choose piano sound and colors; learn to play the piano; customized learning engine; displays animated sheet music that moves in real time, highlighting the appropriate keys of the keyboard; learn and play at the same time; displays the notes visually using falling note mode; (iOS 3.0 or later).

PianoAngel Free (http://bwinnovationtw.blogspot.com/2011/03/pianoangel.html) App is for users already skilled in piano and for beginners; Apple Core MIDI compliant; (iOS 4.0 or later).

Pocket Piano (http://www.betterdaywireless.com/BetterDayWireless/BDW.html) Carry a piano everywhere; rich sound and versatility of a real piano; full keyboard with all eighty-eight keys including sharps and flats; play favorite tunes; hold keys for longer notes or tap for shorter ones; designed exclusively for the iPad to support up to ten keys being played simultaneously; resize keys to exactly fit finger size; realistic piano sound, recorded from the world's finest grand pianos; full multi-touch support for chord and two-handed play; preset keyboard sizes or customize; adjust the width and height of all; play piano on a single keyboard row or play on two rows; customizable metronome; glissando feature; customize piano sound; adjustable reverb setting to get the same realistic effect as the sustain and soft piano pedals; navigation bar for each row; slide navigation bar to quickly move from A0 to C8 or to anyplace on the piano with the tap of a finger; adjustable master volume; sixteen-bit sound; (iOS 3.2 or later).

Real Piano (http://realpianoapp.com) Produces authentic piano sound quality that was sampled from a real grand piano; also plays guitar, bass, harp, marimba, and music box; offers a set of audio effects, a variety of customizable key labels, professional tuning, and transposition functions; recording and sharing features; see notation in real time; touch pedal panel with foot; adjust settings in a separate screen; full eighty-eight-key keyboard that user can slide and pinch to move and zoom; dynamic expression control; different keyboard layouts and customizable key labels: C-D-E, 1-2-3, or do-re-mi; audio effects support; professional tuning and transposition; record and share to Facebook or SoundCloud or with e-mail or iTunes; Game Center achievement support; connects with the Real Piano Remote app; (iOS 6.1 or later).

Real Piano (http://www.rodrigokolb.com.br/#apps) Digital piano with grand piano, electric piano, synth, organ, acoustic guitar, electric guitar, electric bass, synth bass, strings, horns, clav, harpsichord, toy organ, banjo, accordion, sitar, vibraphone, flute, vocals, and sax sounds; play live music; multi-touch; complete keyboard with twenty realistic instruments; studio audio quality; five octaves; record mode; play in loop; rename recordings; works with all screen resolutions; for pianists, keyboardists, musicians, performers, and artists; (Android 2.1 and up).

Touch Piano! (http://bellstandard.com) Featured in more than forty countries; Retina graphics; realistic piano sounds; play along to iTunes library; record and playback sessions; choose from a variety of sound packs; high-quality realistic piano simulation; high-quality graphics and audio; practice piano skills solo or jam with friends; (iOS 6.1 or later).

Virtual Piano (http://www.nullapp.com/details.php?id=4) Select between single or double piano keyboard and different sounds; piano game; compose pieces and learn how to play piano; give piano lessons to others on the same tablet; full multi-touch experience on a single or double keyboard; full-range piano keyboard with six octaves; realistic instruments including grand, electric, and honky-tonk piano; studio sound quality; multiple screen sizes; (Android 2.2 and up).

Virtuoso Piano Pro Classic (Peter Nagy) Seven octaves of sampled concert grand piano sound; master tuning; soft (una corda) and sustain (damper) mode with separately sampled sounds; adjustable keyboard up to four octaves; metronome; automatic tempo conversion; built-in reverb with adjustable mix level; key labels with universal nomenclature and color indication; record, save, and play back with customizable project names; overdubbing for complex recordings; play chords with up to five fingers; slide fingers to roll the keys; (iOS 2.2.1 or later).

PIANO APPS FOR CHILDREN

Abby Musical Puzzle: Kids Animal Piano Toy (http://www.22learn.com/app/10/musical-puzzles.html) Musical toy for babies and toddlers; six activities that foster children's creativity, motor skills, and appreciation of sounds and music; for ages two to eight; (iOS 5.1.1 or later).

Electric Piano for Little Composers (http://littlecomposers.com) Features built-in lessons, songs, and a unique ear training module where children guess the missing note; easy-to-use, intuitive interface; step by step lessons; anyone can play; color-coded play notes that respond to touch; simple songs to encourage memorizing and composing; extensive help pages; two additional hidden keys that are easy to find; videos to demonstrate all features and a support forum on the developer home page; (iOS 4.0 or later).

Kid Songs Piano!—Learn to Read Music (http://visionsencoded.com/fun-iphone-apps) Piano app provides visual effects and teaches nine songs for kids to learn including "Bingo," "Hickory Dickory Dock," "The Farmer in the Dell," "Hot Cross Buns," "Mary Had a Little Lamb," "Muffin Man," "Jesus Loves Me," "Ring Around the Rosie," and "She'll Be Comin' 'Round the Mountain"; pick a song from the rotary dial at the left, then tap the star to see the animation begin; can adjust playback speed by dragging the arrow at the top center of the screen; drag towards the rabbit to speed up the playback or toward the turtle to slow it down or anywhere in between; can change the sound the keyboard makes by tapping the round button on the right-hand side just above the keyboard; tap the gear icon to see the Settings screen; on the Settings screen, the displayed letter notation can be set to C D E F G A B or Do Re Mi Fa Sol La Ti; (iOS 4.3 or later).

Kid-Synth (http://www.xample.ch/kids-synth/) Synthesizer designed for kids; has many features normally found on real synthesizers; polyphonic engine; up to twelve notes can be played at the same time; reverb sound effect; delay/echo sound effect; nine multi-layered samples included in free version; twenty-one additional multi-layered samples available as in-app purchase upgrade; two opposite keyboards to let adult play with child or two kids play together; each keyboard has its individual settings (selected sample, effects level, etc.); available samples include drum kits, acoustic instruments, electronic instruments, and animal sounds; (iOS 3.2 or later).

Kids Animal Piano (http://radlemur.com/animal-piano/) Piano game; edutainment; ad-free; colorful nine-tone piano with free play; slide and swipe finger on piano for comical effect; contains the following songs: "Bingo," "Black Sheep," "Brother John," "Happy Birthday," "This Old Man," "Itsy Bitsy Spider," "Twinkle Twinkle Little Star," "Five Little Monkeys," "Hush Little Baby," "Alphabet Song"; contains animal sounds: dog, cat, rubber duck, frog, chick, pig, sheep, donkey, monkey, horse, lamb, cowbell, piano, violin, cymbal; (iOS 4.0 or later; Android 2.2 and up).

Kids Animal Piano—Preschool Music Game HD (http://tabtale.com) Includes eleven musical note types and nine animals each with their own animal sounding notes; animations, HD illustrations, and sound; for toddlers; has three classic songs that play; while the keys on the piano are pressed, children can learn how to play the songs; drum and classic piano settings; designed to help children enjoy and appreciate different musical instruments; spend musical quality time with child; nine animal music settings include cat, dog, chicken, cow, crow, owl, elephant, goat, and duck; three classic songs include

"Old MacDonald Had a Farm," "Twinkle Twinkle," and "Are You Sleeping?"; musical settings include piano, cat, dog, and chicken; purchasing full version removes all ads; (iOS 5.0 or later).

Kids Halloween Piano (http://radlemur.com/halloween-piano/) Kids piano with a Halloween theme; variety of crazy and fun instruments such as: ghost, crow, witch, organ, scared girl, boiling cauldron, zombie, bell, coffin, water drop, broken music box, and chainsaw; pro version contains more than thirty instruments: knife, tomb opening, mad scientist, violin screech, scared boy, black cat, electric guitar, cricket, chime, cello, whip crack, bulldog, Frankenstein, chains, electricity, double bass, old clock, and door; game comes preloaded with more than ten songs, including the grim "Funeral March" and "Ride of the Valkyries"; contains popular kids songs such as "Bingo," "Black Sheep," "Brother John," "Happy Birthday," "This Old Man," "Itsy Bitsy Spider," "Twinkle Twinkle Little Star," "Five Little Monkeys," "Hush Little Baby," and "The Alphabet Song"; (Android 2.1 and up).

Kids Music Maker (http://www.generamobile.com) Designed to stimulate children's musical development and help them enjoy instruments and songs; includes piano, xylophone, harp, accordion, harmonica, and flute and six children's songs; three different game modes include automatic mode, learning mode, and free mode; (iOS 3.0 or later).

Music for Little Mozarts (http://www.musicforlittlemozarts.com/) Developed for the preschool age group; games provide a balance between learning key aspects of the piano and the enjoyment of making music; learn with Beethoven Bear and Mozart Mouse; (iOS 3.2 or later).

Music Keys (http://www.foriero.com/en/pages/music-keys-page.php) App teaches how to play popular children's songs and recognize piano keys; (iOS 6.0 or later; Android 4.0.3 and up).

Piano Band (http://www.tabtale.com/) Includes eighteen classic children songs, ten catchy musical loops, and eleven musical instrument sounds; three unique game modes include one for playing the song, one for learning the song, and a game challenge; dancing characters move with the pressed piano keys; (iOS 5.0 or later).

Piano Monkey (http://www.brianwestapps.com) Designed to help beginning music students develop quick note recognition in an approachable, game-like environment; animated monkey and cartoon sounds; timed note recognition in four octaves; no sharps or flats; confidence-building multiple-choice format; purple keyboard highlights provide immediate feedback for students; corresponding piano note is also played to help players develop pitch recognition; option to advance questions by shaking the device or using a button; round timing enables parent or child to compete by alternating rounds; game mode setting allows separate treble clef or bass clef practice; cartoon monkey provides visual feedback for kids; (iOS 4.3 or later).

Piano Star!—Learn to Read Music (http://visionsencoded.com/fun-iphone-apps) Piano app provides visual effects and teaches a song for kids to learn; can help beginners learn the letter names for notes on the bass and treble clef including F on the bass through G on the treble; key letters can be set to C D E F G A B or Do Re Mi Fa Sol La Ti; (iOS 4.3 or later).

Pluto Learns Piano (http://pluto-media.com/app/plutolearnspiano/) Player helps Pluto catch the musical notes while avoiding sharks, rocks, and fearsome jellyfish; side-scrolling adventure; play songs mastered on the piano included in the game; current songs include: "Twinkle Twinkle Little Star," *Swan Lake*—Tchaikovsky, "Old MacDonald Had a Farm," *The Magic Flute*—Mozart, "Ode To Joy"—Beethoven, "Oh! Susanna," "The Wheels on the Bus," "Cantata No. 147"—J. S. Bach, "Minuet in A Major"—Boccherini, "Lullaby"—Johannes Brahms, "The Sailor's Hornpipe," "Gymnopédie No. 1"—Erik Satie, and *Sleeping Beauty*—Tchaikovsky; (iOS 4.3 or later).

PIANO LESSONS APPS

50in1 Piano and ***50in1 Piano HD*** (http://www.50in1piano.com/overview/)
Learn to play the piano, create songs, and sing compositions; combines a piano
keyboard, fifty studio-quality instruments, one hundred piano lessons, real-
time effects, one hundred beat loops, and more in a user-friendly interface;
photorealistic keyboard; instant positioning and resizing with gestures; two
keyboard row mode; key labels; eighteen demo songs; microphone recording; four real-time effects; six reverb styles; delay with adjustable timing and
feedback; three-band equalizer; pitch bend controlled via device tilt; Apple
Core MIDI compliant iOS app; allows user to connect MIDI class compliant
hardware to iOS devices via interfaces including the Apple iPad Camera Connection Kit, the Griffin StudioConnect, the IK Multimedia iRig MIDI, the
Yamaha i-MX1, the Line 6 MIDI Mobilizer II, the iConnectivity iConnect-
MIDI, or the Roland Wireless Connect WNA1100-RL; (iOS 3.1 or later).

Glow Piano Lessons (http://appsforhunger.com/Apps_For_Hunger/Home
.html) Learn how to play a song and exercise fingers in a way that prepares user
to play the piano; play the songs in the app or use while playing a real piano;
more than thirty songs; (iOS 4.0 or later).

iLovePiano (http://www.iq-mobile.com/en/iLovePiano) For playing the
piano; learn note names, piano key names, and their location on the staff; includes ear training and basic music theory; make links between a note's sound,
name, position on the keyboard, and position on the staff; learn the musical
alphabet; (iOS 5.1 or later).

Perfect Piano (http://www.revontuletsoft.com/index.html) App to learn to
play the piano with realistic instrument sounds; full keyboards of piano; single
row mode, dual row mode, two player mode; multi-touch; touch pressure
detecting; key width adjustment; six keyboards program: grand piano, bright
piano, music box, organ, Rhodes, and synth; record and playback; supports MIDI and Audio; share recordings; set the recording files as ringtone;
metronome support; Learn to Play mode includes Falling Ball mode, Music
Sheet mode, and seventy preloaded sample songs with more sample songs
downloadable; song library updated every week; support for external MIDI
keyboards over USB; devices should support USB host with Android and
connect with a USB OTG Cable; sound plug-ins include cello, flute, high-
quality grand piano, soprano sax, violin, strings ensemble, electronic piano,
clean electric guitar, and nylon guitar; (Android 2.2 and up).

Perfect Pitch Piano—Ear Trainer (http://perfectpitchpianoapp.com) Ear
trainer game like Simon Says; plays a melody and user plays it back by ear;
teaches user to play piano by ear in a fun and intuitive way; improve ear using the piano; skills will transfer to guitar, singing, drums, etc.; learn to play

popular melodies including "Mary Had a Little Lamb," "Twinkle Twinkle Little Star," "Happy Birthday," "Old MacDonald Had a Farm," "The Farmer in the Dell," "The Wheels on the Bus," "The Ants Go Marching," "Camp Town Ladies," "I Saw Mommy Kissing Santa Claus," "Yankee Doodle," "When the Saints Go Marching In," "Bingo," "Hush Little Baby," and more than thirty more songs available to purchase, including famous classical music from Beethoven and Bach; includes major scales, minor scales, pentatonic, and blues scales; popular folk songs and anthems; beginner, intermediate, and advanced blues riffs; for all levels; more than two hundred individual melodies to master with practice; control melody difficulty, tempo, and pitch; includes scales, intervals, rhythm, theory, pitch, dictation, and more; (iOS 5.0 or later).

Piano 2 (http://www.shiningcode.com/shiningcode/04_en/piano-2.html) Features Key-Marks and Keyboard Sync for teaching music theory; user-friendly design; control keyboard rotation, keyboard size, and range selection in one interface; on-screen instructions; dual keyboards: two full-range keyboards; Key-Marks and Keyboard Sync: innovative features for teaching tunes and music theory; Smart Glissando; subvolume control: both keyboards have an independent sound volume control; easy range-selection; keyboard size control: adjust the keyboard size with three different sizes available; pitch name labels; (iOS 5.0 or later).

Piano Apprentice (http://www.ionaudio.com/products/details/piano-apprentice) Learn to play piano with videos by the Piano Guy, Scott Houston; press Piano Lessons and pick a song to learn; start with the right hand alone, then the left, then put the hands together; if using the Ion Piano Apprentice keyboard, the notes will light up along with Scott's hands; follow the lights and begin playing music instantly; to learn to read notes, select Sheet Music from the home screen and watch a basic tutorial on reading notes; speed controls for controlling the playback tempo so user can learn at own pace; create melodies by selecting Piano Jam; (iOS 4.2 or later).

Piano Ear Training Pro (http://www.themelodymaster.com) Ear trainer to improve aural skills; play the piano by ear; nine different games to become a better musician and achieve high scores in graded theory exams: Higher Lower, Pitch Training, Interval Training, Chord Distinction, Scale Identification, Interval Comparison, Pitch Identification, Melodic Dictation, Chord Progression; authentic and configurable piano including grand piano sounds, multi-touch, glissando, highlighting, note names, and feedback; resizable piano suitable for all devices and tablets; suitable for beginners to advanced musicians; games have different levels; program and focus on specific chords, intervals, progressions, scales, etc.; progress chart; each game has help and tutorials; (iOS 6.0 or later; Android 2.2 and up).

Piano Lesson PianoMan (http://www.yudo.jp/en/pianoman) Can be played with one finger; select and download songs; regulate the speed with ten levels; play slowly to begin and then increase the speed; adjust the accompaniment's sound volume; play against four people via a network connection; preset songs list includes classical pieces; (iOS 4.3 or later).

Piano Life HD—Learn Music Theory and How to Sight Read (M San) Learn to play the piano; innovative feature allows users to copy sheet music by dragging notes directly onto the staff; covers basic piano theory; create piano pieces and rearrange them; drag and drop notes onto the grand staff; snap notes into position; shows correct notes played; (iOS 5.0 or later).

Piano Made Easy (http://www.mcpiano.com) The Mayron Cole Piano Method; classical piano lessons for kindergarten to college; includes orchestrated accompaniments, animated illustrations, and interactive theory sheets; each textbook is divided into lessons for students who want to teach themselves to play the piano or for students who are taking lessons from a piano teacher; place tablet on piano as with any music book; for beginners to adults; (iOS 3.2 or later).

Piano Teacher (http://pianisthd.com) Play favorite songs; piano tutor app has 128 instruments and more than fifty thousand songs in songbook; app to learn how to play music on the go; pick a song and learn how to play it; three keyboard themes; recording/social sharing functions; ability to change size of the white keys; double keyboard mode; can load any MIDI file (.mid or .MIDI) and KARAOKE format (.kar) file to play; in-app tutorial; localized to: default, Japanese, German, Chinese, Vietnamese, Korean, Romanian, French, Spanish, Italian, and Russian; Auto Hand feature in practice mode, perform mode, mirror and classic double keyboards modes; in-app built-in video tutorial; instrument sounds include piano hq, hand drum, guitar hq, harmonica, bell, music box, organ, and guitar bass; key-by-key guide mode; key press graphic effect; realistic feeling with vibration; create, share, and download from cloud sharing services; sound options include real sound and digital sound; keyboard and speed control options; two-row and two-row mirror mode with multiplayers and multi-instruments function; single learn to play, single perform, three-row mode; record and playback to create and share with friends and family; learn-to-play mode with the pre-loaded collection of famous songs; kid mode available; various works of great artists available; musical instruments to unlock include piano, drum, trumpet, violin, guitar, bass, and sax; professional feedback and support system via e-mail and help desk; studio-quality sound, MIDI effect and note velocity, aftertouch, and pitch bend; looped playback; free music library from cloud sharing service and public search engine; marketplace for songs and artists; (Android 2.2 and up).

Piano Tutor (http://www.smileyapps.com) Practice piano skills anywhere; for piano beginners, one of the most critical skills is learning how to read the piano sheet; Apple Core MIDI compliant; (iOS 6.0 or later).

Play Piano HD (http://www.mobilesort.com/play_piano.html) Learn how to play piano notes with three different modes: Practice, Points Mode, and Time Trial; grand piano sound lets users hear the notes as they play; use note labels to know which note is which, then gradually turn them off; see progress over time; three levels to ramp up the difficulty over time including easy: simple notes, including B♭ and F♯, medium: more complex notes, and hard: random jumps between complex notes; practice on the go or at home; (iOS 5.0 or later).

Synthesia (http://www.synthesiagame.com) Learn to play the piano using falling notes; play songs on the iPad touch screen or connect a MIDI keyboard; put device on the music stand and play using real keys; Apple Core MIDI compliant; (iOS 5.0 or later).

Virtuoso Piano Free 2 HD (https://www.YouTube.com/view_play_list?p=ADE021B190FF35C4) Learn and play the piano; learn the basics of music; tuning, adjustable reverb, advanced metronome, recording, and more features; six octaves of sampled concert grand piano; robust bass, warm middles, and crisp highs; play along with music library; key labels with colors; play chords with up to five fingers; slide fingers to roll the keys; (iOS 3.1 or later).

Virtuoso Piano Free 3 (Peter Nagy) Learn and play the piano anytime and anywhere; with TrueVelocity hard taps are louder and soft taps are softer; play wirelessly from device to Apple TV and AirPlay-enabled speakers; two pianos included: a classic concert grand and a broken pub piano; eighty-eight keys of sampled pianos; robust bass, warm middles, and crisp highs; adjustable sustain; optional double keyboard; duet mode; play along with music library; key labels with colors; soft mode; AirPlay support; play chords with all fingers; slide fingers to roll the keys; optimized for Retina displays; (iOS 4.0 or later).

PIANO NOTES, CHORDS, AND SCALES APPS

Chordinary (http://www.appsinteractive.com) Includes seventy-eight variations for each chord in twelve keys, plus inversions for each for a total of 936 chords and 564 scales; includes major chords, minor chords, diminished chords, and augmented chords; chord sounds; create chords; e-mail and share chords; (iOS 3.1 or later).

Chordipedia (http://www.appsinteractive.com) Learn and play chords and scales; for beginner to advanced piano and keyboard players; fast response time; individually sampled notes; keyboard range from C2 to C♯5; easy to scroll keyboard; can be used as a stand-alone piano; can be used to play chords with ease; easy chord and scale selection; selected chords are played and highlighted; selected scales are played and highlighted; replay button to play last selected chord or scale; circle of fourths/fifths; jazz and gospel chords; ear training; (iOS 3.1 or later).

FastChords HD (http://www.smartutils.com/) Map chord symbols to piano keys with calculator-like interface; for learning new songs from fake books or from music scores; play own chord progressions; piano chords reference; (iOS 6.1 or later).

Nota (http://notaapp.com) Set of indispensable tools for musicians at any level; piano chord and scale browser; piano and staff note locator; note quiz and reference library with more than one hundred symbols; covers the basics of music notation with a four-octave piano that displays the notes on a staff; includes a full-screen landscape mode piano for practicing and an interactive notes quiz to test knowledge of notes; scales browser has a comprehensive list of common and exotic scales; app will show the scale, play it, and display the notes, intervals, and half-steps; Chords Browser makes it easy to find a chord and play it on any key or invert it; can set the notation to strict or simplified and also set the root to sharp or flat; consult the Circle of Fifths in the Reference section with a comprehensive reference of music notation; Scales Browser includes Major to Minor to Myxolidian and allows user to look up almost any scale, play it in any key, and view its notes, intervals, and half-steps; Quiz Basics note quiz tests knowledge of notes in the staff; difficulty level can be set to easy with thirty-four notes or advanced with eighty-two notes that include sharps and flats; Music Brainiac includes an extensive reference library with more than one hundred items including accents and accidentals, lines, breaks and clefs, key signatures, chords, circle of fifths, dynamics, note relationships, notes, notes and duration, repetition and codas, rest and durations, time signatures, browse with a flick; in fixed Do Solfège mode, notes are shown in their Solfège syllables: Do, Re, Mi, Fa, Sol, La, Ti, Do; tool for anyone doing music, beginner or experienced; (iOS 4.0 or later).

Note Lookup! Learn to Read Music (http://visionsencoded.com/fun-iphone-apps) Learn the letter name of a note and where that note is on the piano keyboard; learn what a piano note looks like on the music staff; switch key signatures from four flats through to four sharps; letter notation can be set to: C D E F G A B or Do Re Mi Fa Sol La Ti; (iOS 4.3 or later).

Notes!—Learn to Read Music (http://visionsencoded.com/fun-iphone-apps) Basic flashcard app; helps beginners learn the letter names for notes on the bass and treble clef including F on the bass through to G on the treble; displays a random note on the musical staff and user taps the correct letter/key; MIDI enabled; use Apple Core MIDI compliant devices with the app; change to either treble or bass clef by tapping the clef; toggle the visual effects on or off by tapping the trophy icon; practice only specific notes; tap the gears icon to enter the Settings screen to disable all but the specific notes needed to practice; (iOS 4.3 or later).

PChord (http://www.asoft.ne.jp/iOS/en/) Chord book for piano; (iOS 5.1 or later).

Piano Chords and Scales (http://movilcrunch.com/piano-chords-and-scales/) Reference for piano students and pianists; teaches different chord and scale combinations; select a chord or scale and check it on the keyboard; chords and scales for the twelve keys; 126 chord combinations for each key; 108 scale combinations for each key; selected scales and chords are displayed on a keyboard; hear how scales and chords sound; (iOS 4.3 or later).

Piano Notes Fun (http://www.shiningcode.com/shiningcode/03_en/piano-notes-iphone.html) Practice reading music notes; wide range of level control; fun game for experienced piano players; music note tutor for beginning piano students; learn and practice music notes effectively; large note range: covers more than four octaves from B0 to D5; Teaching Mode: hitting a key shows its note and pitch name; Game Mode: practice sight-reading skills with games; range selection: adjustable range lets user practice particular notes intensively; all key signatures: fifteen key signatures for treble clef and bass clef respectively; help modes: three help modes, including a tutor mode for beginners; colored keys: while practicing, colors can be applied as hints for correct keys; pitch name labels: white keys can be marked with their pitch names; multiple notes: shows multiple notes simultaneously; game length control: each game quizzes user with ten pages of notes and the number of notes is adjustable; high-score records: every treble and bass scale has its own score; high-quality sound; (iOS 5.0 or later).

Piano Notes Pro—Sight Reading Tutor (http://www.shiningcode.com/shiningcode/02_en/piano-notes-ipad.html) Practice reading music notes; aid for piano teachers; wide range of level control; fun game for experienced piano players; music note tutor for beginning piano students; learn and practice music notes effectively; full-range keyboard covers all eighty-eight keys of a piano keyboard; Teaching Mode: hitting a key shows its note and pitch name; Game Mode: practice sight-reading skills with games; range selection:

adjustable range lets user practice particular notes intensively; quiz types: four quiz types including random, ascending, descending, and chords help improve skills in different ways; additional quiz types available via in-app purchase; all key signatures: fifteen key signatures for treble clef and bass clef respectively; help modes: three help modes, including a tutor mode for beginners; colored keys: while practicing, colors can be applied as hints for correct keys; pitch name labels: white keys can be marked with their pitch names; multiple notes: shows multiple notes simultaneously to help improve reading speed; game length control: each game quizzes user with up to ten pages of notes with ability to adjust the number of notes and the number of pages; achievements; high-score records: scores of all key signatures of both clefs are listed in a table as a progress report; keyboard size control: can adjust the keyboard to three different sizes; multiple user profiles: up to forty user profiles; share app with friends or family members; useful for piano teachers to keep track of students' progress; high-quality piano sound; MIDI Input: supports MIDI input with Core MIDI via Apple Camera Connection Kit and USB port on keyboard; (iOS 5.0 or later).

Piano Notes! Learn to Read (http://visionsencoded.com/fun-iphone-apps/) Basic flashcard game; helps intermediate learners to associate notes on the bass and treble clef in the keys of C, F, G, B♭, D, E♭, A, A♭, and E (zero to four flats, one to four sharps) with the correct piano keys; challenging Game Modes include Arcade, Count-down, and Endless; (iOS 3.1 or later).

Piano Scales and Jam Pro (http://www.learntomaster.co.uk) Learn scales, modes, and broken chords in any key on the piano with thousands of combinations; listen, play, and check scales or view as a reference; highlights the right notes to play in jam mode; configurable piano includes multi-touch, glissando, highlighting, and note names (standard or Solfége); resizable keyboard suitable for all devices and tablets; learn how to improvise and solo within a chosen key, mode, or chord; set the size of the keys on the piano; the larger keys make it easier to get the right notes; the smaller keys show all three octaves on the screen; full support for all phones and all tablets; authentic digitized grand piano sounds for each separate key on the piano; highlight notes on or off to play by ear; slow down or speed up the scale; ascending and descending scales; show or hide the note names; record and save favorite melodies; highlights the whole keyboard to show which notes are within a chosen key, mode, or chord; thousands of scales, keys, modes, and chords to choose from; play along to a metronome after adjusting percussion backing, bpm, type of beat, and metronome volume; select a scale from the extensive list; listen and watch the notes being played through highlights and then play them back; check answers; start with a few notes and work up to more until mastered; select the root of the scale and type of scale or key, mode, and broken chord from hundreds of scales including popular ones (major, minor, harmonic minor, pentatonic, blues, etc.) and rare ones (altered, gypsy, Persian, etc.), all the modes (aeolian, dorian, flamenco, etc.), and all the broken chords (7(b5), Diminished, 6 add 9, etc.); (iOS 6.0; Android 2.2 and up).

Piano Tabs (http://pianotabsapp.com) Helps user play, remember, and compose songs on the piano; create new songs with lyrics and chords and display them ready to play; look up chords in the chords section; use the search feature to find the chord currently playing on the piano; play, compose, and sing songs; (iOS 4.3 or later).

PianoHead (http://spinapse.com/) Solution for mastering the fundamentals of music theory; set of modular drill-and-practice games designed to build automaticity for note recognition, scales, intervals, and key signatures; nine levels in the interactive game; (iOS 3.2 or later).

PIANO REPERTOIRE AND POPULAR SONGS APPS

Classical Piano (iPlayTones, LLC) Award-winning pianist James Lent brings some of the best classical piano music and ringtones from the U.S. cellular networks directly to all devices; includes over fifty-three minutes of classical piano music and twenty ringtones; play all fourteen full-track MP3s and twenty ringtones; download in bundles or individually; includes works by Franz Joseph Haydn, Johann Pachelbel, J. S. Bach, Ludwig van Beethoven, and Wolfgang Amadeus Mozart; (iOS 5.0 or later).

Contemporary Piano (iPlayTones, LLC) Contemporary piano music and ringtones from the U.S. cellular networks by award-winning pianist James Lent; featured composers include Samuel Barber, Aaron Copland, George Gershwin, Aram Khachaturian, Gyorgy Ligeti, Nicolai Rimsky-Korsakov, Frederic Rzewski, and Erik Satie; includes fourteen full-track MP3s and twenty ringtones; download in bundles or individually; (iOS 5.0 or later).

ezPiano for iPad (http://www.innovaecanada.com/products/ezPiano-for-iPad) App that works like *Guitar Hero* for the piano; shows user the sheet music and plays the accompaniment for them while they are playing the tune; all one hundred songs come with a full accompaniment; play a selected melody note by note and the app will fill in the harmony and accompaniment; turn off this feature to play just the melody; choose songs from Broadway, Classical, Folk Songs, Impress Your Friends, Kids, and Miscellaneous; includes scales; stereo piano sounds sampled from a Steinway Model D Concert Grand piano; learn how to read the sheet music by pressing a key and watch the corresponding note in the sheet music highlighted; learn how to play a song the correct way by following the fingering guide; check score to gauge accuracy; includes a metronome; Song Mode: learn how to play a song or let the app play it; Free Mode: play your own song; Random Mode: let the app play a random tune; Fake Mode: pretend to play any song like a pro; sound effects when user makes a mistake or does an excellent job; comical facial expressions in reaction to how well user is playing; animations; make the keyboard slide to change octaves quickly; can turn off the fingering and note guides; press to see all the settings that can be configured; (iOS 4.2 or later).

Logical Piano (http://www.appscraft.ru/logicalpiano/) A game in which user rehearses eighteen popular melodies by repeating keystrokes one by one; songs range from "Twinkle Twinkle Little Star" to a Beethoven masterpiece; number of tunes will increase in updates; HD version available; game has been translated into ten languages; develops memory and sense of rhythm; for children and adults; can open the lid of the piano; (iOS 4.3.3 or later).

Magic Piano (http://www.smule.com/apps#magic-piano) Includes songs from Bruno Mars to Mozart; available in HD; offers free songs every day and includes a large catalog of songs; turn on game mode to unlock achievements and free songs; more than twenty-five million players; make music and share performances; more than one thousand songs with new hit songs added every week; earn new songs by playing and by watching videos; songs will sync across multiple devices; current songs include pop, rock, classical, movies and musicals, country, and RnB hits; suggest songs on Smule's Facebook page; follow beams of light to guide fingertips to the correct notes; control the notes, rhythm, and tempo; try out different instruments; change piano into a harpsichord, funky 1980s synth, organ, and more; enter game mode and get rewarded for playing; level up to unlock achievements and badges; earn free songs and get to the top of the song leader boards; change the difficulty of each song to give a unique challenge; broadcast performances on the Smule Globe or listen to other players' songs; share pieces through Facebook, Twitter, and e-mail; subscribe for unlimited access to full song and instrument catalog; play all available songs on any instrument for the duration of subscription; (iOS 6.0 or later; Android 4.0 and up).

Master Piano (https://itunes.apple.com/us/app/master-piano/id364897373 ?mt=8&ign-mpt=uo%3D4) More than 250 free songs included; record own songs; learn a song at the tempo the song is played with animated falling notes; resize keys for a custom feel; concert hall sound; innovative songwriting and learning capabilities; technical innovations written exclusively for the iPad; realistic piano sound, recorded from the world's finest grand pianos; eighty-eight keys including all sharps and flats; full multi-touch support for chord and single note play; easy-to-use preset keyboard sizes or customize own; adjust the width and height of all keys; play on a single keyboard row or play on two rows with dual row; identify notes with optional note labels; easy to use, fully-customizable optional metronome; glissando feature allows user to slide finger across the keyboard like a real piano; adjustable reverb setting to get the same realistic effect as the sustain and soft piano pedals; navigate the full piano keyboard with a unique navigation bar for each row; slide navigation bar to quickly move from A0 all the way to C8 or jump to any place on the keyboard with the tap of a finger; customize each navigation bar's highlight color using preset options or create own unique color; adjustable master volume; record, edit, and save songs; record all nuances of performances including duration of note and tempo; no time limit on the length of recordings; overlay unlimited tracks onto recordings (overdub); save compositions for later into "My Songs" folder; all songs can be loaded for playback, editing, or learning; adjust the tempo of any song; supports "Learn Song" mode for compositions, so can play them back and learn them note by note; edit compositions to fine tune; Song Learning features more than 250 preloaded songs in fifteen different categories including children's, classical, Christmas, ragtime, folk, and world music; songs are divided into twenty difficulty levels; difficult songs feature chords and include both left and right hands; Favorites folder to add and remove songs; change the tempo of any song to match playing level; shift the song's octave up or down; learn the whole song or a favorite part; all learning modes teach notes, tempo, and note duration; Portion mode to focus on learning left or right hand for a piece; auto-scrolling moves the on-screen keyboard while playing the notes; (iOS 3.2 or later).

New Age Piano (http://www.anywhereartist.com/aa/Welcome.html) Collection of relaxing and inspiring melodies; features music, news, photos, videos, updates, and more from twenty New Age pianists; often referred to as Contemporary or Neoclassical piano, New Age piano music is music of various styles intended to create artistic inspiration, relaxation, and optimism; used by listeners for yoga, massage, meditation, and reading as a method of stress management or to create a peaceful atmosphere in the home or other environments; emphasizes artistic and aesthetic expression; strongly influenced by and sometimes also based upon early baroque or classical music, especially in terms of melody and composition; artist may offer a modern arrangement of a work by an established composer or combine elements from classical styles with modern elements to produce original compositions; many artists within this subgenre are classically trained musicians; Neoclassical New Age music is generally melodic, harmonic, and completely instrumental; (iOS 5.0 or later).

Piano Complete (http://www.betterdaywireless.com/BetterDayWireless/BDW.html) Complete piano with the ability to learn songs; more than five hundred songs in a variety of categories; learn both hands or just the right hand or left hand; eighty-eight-key piano with multi-touch, adjustable reverb, metronome, dual row, completely resizable keys, high-quality sixteen-bit piano sound, chord support, glissando, sizable sharps, navigation bar, note labels, theme colors, sustain pedal, and adjustable metronome; learn songs with falling notes; sheet music; (iOS 3.0 or later).

Piano Dust Buster 2 (http://www.joytunes.com/piano/index.php) Popular app in more than eighty countries; more than forty million songs played; use own acoustic piano or an on-screen virtual piano to play favorite pop or classical songs; award-winning music game activated by an acoustic piano/keyboard or by touch; no wires or adapters needed; app listens to piano and user learns how to play; two playing modes to choose from for both players who know how to read music or not; compete against friends and players from around the world; in Concert Mode there is always a competition to compete in and free content to play; win medals, level up, and unlock even more songs from the JoyTunes library; in Jukebox Mode can practice piano skills while playing fun pop songs and classics; gain XP (experience) points and level up to unlock more new songs; purchase unlimited access to song library or individual song packs of popular and fun songs for all different skill levels, accompanied by high-quality concert music; new pack of songs added every month; various daily free offers; more than 150 songs from different genres available; virtual piano has real-size 3-D keys in iPad version; learn note names, rhythm, playing songs, and more; special Staff Mode for practicing sheet music notation; (iOS 6.0 or later).

Piano Free with Songs (http://www.betterdaywireless.com/BetterDay Wireless/BDW.html) Offers a large song catalog with more than eight hundred songs to choose from; perform songs in the style of artists such as Bruno Mars, Pink, Katy Perry, Adele, and Lady Gaga; songs from every genre including current hits, popular children's songs, and classical pieces; eighty-eight-key piano with multi-touch, adjustable reverb, metronome, dual row, completely resizable keys, real piano sound, chord support, glissando, sizable sharps, navigation bar, note labels, theme colors, sustain pedal, and adjustable metronome; learn to play songs; new songs added every week; songs available on all devices; includes a traditional piano and the interactive Starfall piano; high-fidelity, professionally recorded sounds; perform any song with a band; real-time experience; licensed and legal; unlimited premium subscription; (iOS 4.3 or later).

Piano Genius (http://www.synapticstuff.com/piano-genius-ipad) Piano game; play favorite songs; more than four hundred songs from Mozart to Coldplay; track progress, compete with others, and share scores with friends while learning actual piano skills; intelligent accompaniment shows user when and where to play notes and chords while automatically playing the rest of the song; no experience required; new songs added every week with free songs and weekly contest rewards; current songs include Modern Hits, All Time Classics, Rock, Classical, and more; falling dots show which keys to hit in which order, dynamically playing the rest of the notes; keyboard automatically scrolls to position for the next note; to top the score charts user needs to master timing, duration, and anticipate bonus notes shown on the notation view; unlock free songs and in-app features by earning achievements and hitting objectives; multiple levels of song difficulty (easy, medium, and hard); earn achievement badges to share on Facebook or Twitter directly from the app; compare song performance with the rest of the Piano Genius community; leader boards for every song; earn free credits through weekly tournaments; challenge friends; retro virtual piano with MIDI connection capability to play freestyle with digital concert piano sound, scrollable full-octave keyboard, and notation feedback to show which keys correspond with which notes; new upgrades added every month; (iOS 5.0 or later).

Piano Mania (http://www.joytunes.com/piano/pianomania.php) Piano practice game featuring hundreds of songs; fun music journey while playing own acoustic or electric piano; no wires or adapters needed; place it on your piano/keyboard and play away; touch mode also available; makes practice time fun and effective; master each chapter to open the next, all while improving sight-reading, rhythm, technique, playing two hands, and much more; keep progress synced on any iPad; makes sheet music come to life; learn to read sheet music notation and symbols; play melodies in both treble and bass clefs; choose between right hand, left hand, or both hands simultaneously; progress up in chapters, collecting skill points; new songs and levels added every month; library includes beginners' classics, popular hits, and hundreds more; all app users can play several free skill levels; upgrade to a subscription package at any time; complementary tool to piano lessons; progress in levels, earn ranks, and enjoy full access to song library; new songs added weekly; subscription options; (iOS 6.0 or later).

Piano Melody Pro—Learn Songs and Play by Ear (http://www.learntomaster .co.uk) Learn to play songs on the piano by playing back the melody; four hundred songs to learn from different eras and genres; configurable piano includes multi-touch, glissando, highlighting, and note labels; resizable piano suitable for all devices and tablets; set the size of the keys on the piano; focus on parts of the song to play back and learn the licks; authentic digitized grand piano sounds; highlight notes on or off to play by ear; show or hide the note names; slow down or speed up the song; turn up or down the volume; develop ability to play by ear; suitable for all ages, kids to adults, and for all abilities, beginner to advanced; song list is varied containing all genres such as Rock, Classical, The Latest Pop, Film Theme Tunes, TV Theme Tunes, 60s, 70s, 80s, 90s, Naughties, Modern, Alternative, Indie, Latin, and more; songs capture the main melody, chorus, intro, and verse and contain up to four hundred notes; uses standard note tabs where lowercase are the white notes and uppercase the black notes; (iOS 6.0 or later; Android 2.2 and up).

Piano Music Collection Pro (http://strongwang2012.blog.163.com) Music player with knowledge; includes portraits of the composers; read about the composers while listening; high-quality music; shuffle or order; repeat mode; list music by composers or by favorites; search by composer or music; background play; includes twenty great composers and their ninety-nine best piano pieces including Bach, Mozart, Liszt, Beethoven, Chopin, Satie, Scarlatti, Debussy, Ravel, Rachmaninoff, Mendelssohn, Mussorgsky, Schubert, Schumann, Prokofiev, Scriabin, Grieg, Pärt, Tchaikovsky, and Balakirev; (iOS 4.3 or later).

Piano Star (https://play.google.com/store/apps/details?id=com.gameclassic .musicstar) Challenging piano game to practice music rhythm and reaction speed; new music is added to the app regularly including Pop, Rock, Classical, Movies, TV, Reminiscence, Game, Songs for Children, and more; world music contest; Global PK Mode and Ranking system; earn golden coins to buy locked songs; lighting effects and animations; (Android 2.1 and up).

Piano Summer Games—Play National Anthems (http://www.joytunes.com/ summergames/index.php) Represent your country and compete for a medal against other players from around the world; learn to play the anthems of the world on own real piano or an on-screen virtual one; no wires or adapters needed; app listens to piano and user learns how to play; growing list of more than forty national anthems; learn how to play them accompanied by high-quality concert music; in Competition Mode users compete in real time with players from all over the world in playing an anthem; get into the top three to win a gold, silver, or bronze medal; worldwide piano Olympics; practice and learn songs, then compete with people from other countries for a place on the podium; learn to play almost every national anthem from all over the world; live press feed on competitions and winners; virtual piano has real-size 3-D keys (iPad version only); works on musical skills; learn note names, rhythm, playing songs, and more; special mode for practicing sheet music notation; automated download of new songs; (iOS 5.0 or later).

Player Piano Free (http://www.betterdaywireless.com/BetterDayWireless/ BDW.html) Plays piano music and lets user play it too; play the top row and let the player piano play the bottom row; play the bottom row and let the player piano play the top row; try to play both rows; songs are the complete song with chords and both hands; can choose how much or how little of the song to play and let the player piano play the rest; displays the notes falling in smooth animation to the tempo of the song; keys are highlighted to let user know to press them; each piano note is accurately labeled and has a tail to further signify how long the note should be held; many categories and songs; (iOS 3.0 or later).

PlayItYourself 4 HD (http://alfaproductions.com) Game that helps user learn to play the piano; score and upcoming notes are highlighted for user to play on the built-in piano; can study scores or follow the notes shown by the app; choose what to practice and the app will play the rest; edit and export own music with free MuseScore desktop app; sync scores through iTunes; shows authentic, professionally engraved sheet music reformatted for optimal viewing on mobile device; full-size multi-touch piano is built-in; high-quality sound samples; choose the hand to play, and the other hand will be played by the app automatically so user can concentrate on the melody; tempo of the second hand is adjusted automatically to match the speed; includes "Fur Elise," "Moonlight Sonata," "Air on the G String," "The St. Louis Rag," "Ladoiska," "La Pastorale," "Les Graces," "Tambourin," "Pictures at an

Exhibition (Promenade)," "The Entertainer," "Piano Sonata No. 11," "Rondo Alla Turca," "Chopin Waltz," "Waltz in A Flat," "In the Hall of the Mountain King," "The Star Spangled Banner," "Spring," "Carmen Prelude," and "Aria in D"; (iOS 3.2 or later).

Pocket Jamz Piano Notes (http://www.synapticstuff.com/support/piano-notes-ios) Learn to play favorite artist's songs; all-in-one piano learning and reference tool; download, upload, or purchase fully interactive sheet music with adjustable tempo, multiple-track playback, audio playback, and fingering guides that show exactly which keys to press; context-sensitive effects legends show user how to play any effects present in a song; dynamic loop points allow user to repeat difficult sections; intelligent song library automatically groups related arrangements and includes YouTube videos showing how to play songs; Wikipedia background on song history; high-quality audio playback; interactive music scores with selectable tracks; hands-free scrolling playback with adjustable song tempo and metronome; connects to the AirTurn Bluetooth page turner; included with each download is a starter pack of songs from composers like Bach, Beethoven, and Pachelbel; Synaptic Stuff has partnered with top music publishers to offer high-quality, interactive editions of authentic song arrangements including lyrics; more than two hundred thousand community songs; purchase an in-app subscription to access huge catalog of community-generated songs; rate, review, and interact with the Piano Notes community using song-specific threads; select songs and pieces from classical, pop, jazz and blues, Christian and gospel, classic rock, children's songs, country, metal, reggae, RnB, and soul; set loop points anywhere in the song to automatically repeat difficult measures; dynamic tempo settings; click on any compatible song from mobile web browser or e-mail attachment and have it automatically imported into the application for storage, access, and playback; upload piano scores in MusicXML format using WiFi transfer, web, or e-mail; full-page realistic sheet music rendering on the iPad; beginner, intermediate, and advanced arrangements; (iOS 6.0 or later).

Pocket Piano Song Universe (http://www.betterdaywireless.com/BetterDay Wireless/BDW.html) Learn hundreds of songs on the piano by playing in time with the falling notes on the screen; feature-loaded piano that includes the ability to upload and download songs from a global community; unique piano training mode that will teach user how to play any song in the library including Beethoven to modern-day hits; song library includes some of the most popular songs, themes, and movie soundtracks of all time with more than one thousand songs; (iOS 3.0 or later).

Power Piano (http://pianogenius.duapp.com/pianojoy2/) Innovative app with sound effects; play melodies with a drum, flute, guitar, and other musical instrument sounds; more than one hundred classic, romantic, and pop piano songs; six piano keyboards; (iOS 4.3 or later; Android 2.2 and up).

Tiny Piano (http://www.tinypiano.com) Touch anywhere on the piano to play the next note in the song; can play fast or slow; comes with more than one hundred free songs, including: "Amazing Grace," "Call Me Maybe," "The Entertainer," "Fur Elise," "Gangnam Style," "Heart and Soul," "Jingle Bells," "Minecraft: Sweden," "Minuet in G," "Never Gonna Give You Up," "Ode to Joy," "Party Rock Anthem," "Skyfall," "Somebody That I Used to Know," "Stand By Me," "Thrift Shop," "What Makes You Beautiful," and "You Raise Me Up"; play songs from artists like Justin Bieber, One Direction, Katy Perry, Bruno Mars, Adele, Lady Gaga, Rihanna, Celine Dion, and Billy Joel; (iOS 5.0 or later).

VARIOUS KEYBOARD INSTRUMENTS APPS

Accordio Pro (http://www.livingtransient.com) Chromatic button accordion; sounds like a traditional punchy Italian accordion; full six-row Stradella bass section; scroll and zoom the keyboards; short videos on website or YouTube; (iOS 3.2 or later).

Austrian Accordion (http://littlecomposers.com) Professional version of the Austrian Button Accordion; beautiful accordion sound; Hammond organ to play pop songs; five bass buttons for playing the accompaniment in C, F, B♭, G, and D; four musical styles including waltz, polka, rock, and ballad; (iOS 3.2 or later).

Galileo Organ (http://www.yonac.com/galileo/index.html) Professional tonewheel and transistor modeling organ for iPad and iPad Mini; eleven unique organ types; comes with three manuals, scanner vibrato/chorus emulation, percussion module, and settable organ parameters such as leakage, key-click, attack and release, and more; signal is rounded out by a virtual-tube, Class-A-inspired preamp, as well as a powerful rotary cabinet simulator with three cabinet types and multiple parameters; performance and studio-friendly; features comprehensive MIDI implementation with more than 130 CC destinations; Audiobus input and effects; built-in arpeggiator, tapedeck, and more; more than 240 factory presets; classic organ sounds; (iOS 5.1 or later).

Harpsichord (Paul Hudson) Hear the harpsichord come to life on device; beautiful sound across four octaves; full support for multi-touch; can play with both hands across two keyboards in the iPad version; use glissando mode to slide fingers freely across the keyboard or enable scrolling mode to have the full range of the keyboard; choose between the classic white on black theme or the two alternatives: wooden and piano; works on iPhone, iPod Touch, and iPad; play the harpsichord anywhere; (iOS 4.2 or later).

Organ+ (http://www.yonac.com/organplus/index.html) Organ emulator with seven different organ engines that re-create the sounds of transistor and tonewheel organs; FX stack with dual chorus/vibrato, vintage reverb, overdrive, and more lets user create variations and hone in on the organ sound; two manuals are individually adjustable with nine drawbars each; distinct tones improve phrasing and effect; can set organ key clicks and percussion; can save presets, record performances, and watch them play back with a phantom pressing the keys; bounce recordings to audio and export them via WiFi; comes with thirty-six presets spanning decades' worth of classic organ tunes; can play using external MIDI controllers; use the appropriate adapter to connect to device's charging port; YouTube channel shows features; (iOS 4.2 or later).

Organist (http://moocowmusic.com/OtherApps/index.html) Play and record electric or pipe organ; single or dual keyboards to suit playing style; optional key labels aid navigation around the keyboard; (iOS 3.0 or later).

Piano Accordio Pro (http://www.livingtransient.com) Piano accordion; fits into the iPad's screen; sounds like a traditional punchy Italian accordion; responsive and fun to play; full six-row Stradella bass section; scroll and zoom the keyboards and play along to music in music library; short videos on website or YouTube; (iOS 3.2 or later).

Pocket Organ C3B3 (http://www.insideout.co.jp/apps/2014/03/25/what-is-the-pocket-organ-c3b3/) Musical keyboard application that simulates the Hammond organ; sound is generated with the synthesis of the waveforms from "Virtual Tonewheel System" controlled by nine drawbars like the actual Hammond organ; Apple Core MIDI compliant; (iOS 3.1.3 or later).

Traditional Musical Instrument Apps (http://www.tradlessons.com/apps.html) Apps enable accordion, concertina, and bagpipes players to learn and practice anywhere; (iOS varies by app).

Virtuoso Piano Celesta (https://sites.google.com/site/virtuosoiphone/) Four octaves of sampled celesta; automatic time stretch and fully configurable keyboard; adjust the reverb level to change the acoustical environment; slide fingers to roll the keys; sustain mode with separately sampled sounds; adjustable keyboard size up to four octaves; built-in metronome; automatic tempo conversion; key labels with color indication; record, save, and play back with customizable project names; play chords with up to five fingers; (iOS 2.2.1 or later).

· 11 ·

Guitar and Bass Guitar Apps

BASS GUITAR APPS

Bass Guitar Companion (http://www.infinautgames.com/apps-for-musicians/ bass-guitar-companion/) Several tools in one; bass guitar tuner; large lists of chord charts and scales; learn how to play the bass guitar; simple intuitive interface; learn to tune by ear by clicking a peg to start or stop; more than 1,600 chord charts show multiple positions for each chord; more than 150 scales from blues and basics to more obscure charts; display options include finger positions, intervals, or notes; auto strum to hear the selected chord; manual strum to play a chord by swiping across the strings; dynamic string indicators show which string or note is playing; metronome plays basic beats with time signature options; variety of drum tracks available; landscape mode for iPad; localized for sixteen languages; (iOS 3.0 or later).

Bass Guitar Tutor Pro (http://www.learntomaster.co.uk) Learn to play favorite bass lines on the bass guitar with more than one hundred songs; learn scales and jam in any key; save riffs; learn typical bass patterns such as reggae, walking bass, and hard rock in any key; resizable fretboard; fingered or slap bass; for beginners to top musicians; high simulator quality; quick and responsive; focus on sections of the song to master bass lines, patterns, and scales; adjust the speed and volume for playing along; highlight notes on or off to play by ear; includes all genres: rock, indie, 60s, 70s, 80s, 90s, naughties, modern, Latin, classical, film themes, TV themes, and traditional; ever expanding catalog of bass lines; select a scale from the extensive list; listen and watch notes being played through highlights and play back; check answers; start with a few notes and work up to the whole scale; change the position; repeat scales ascending or descending; view scale patterns as a reference; record and save favorite jam sessions and play back; play along to a click; provide the bass in any key for any genre: alternative rock, barrelhouse, blues, boogie woogie, bossa nova, country, 1950s rock, funk, hard rock, jazz, and more; (iOS 5.0 or later).

Bassist (http://moocowmusic.com/OtherApps/index.html) Play bass guitar straight from device; experiment with new riffs and learn about the bass; play slap bass or finger-pick; both four- and five-string basses are provided; virtual fretboard can be scrolled up to the twelfth fret; supports left-handed players, flipping over the interface and restringing appropriately; play through headphones or external speakers; (iOS 3.0 or later).

iFretless Bass (http://ifretless.com) Professional virtual instrument that provides bass and guitar players with an expressive fretless playing surface; layout is like a nine-string bass guitar with the notes colored black and white like the keys of a piano; can measure the force of touches on the screen allowing it to respond to dynamics and accented notes with a wide range of different tones for each instrument; more than 200MB of bass samples; MIDI input and output; Audiobus support; Audio copy support; Inter-App Audio support; four-band EQ tuned for maximum control of bass sounds; bass sounds include fretless bass, slap bass, jazz bass, acoustic upright bass, pick bass, electric upright bass, saw wave analog synth bass, chameleon synth bass, snarling synth bass, and FM synth bass; (iOS 7.0 or later).

iJ Bass HD (http://sardineapps.com) Bass app enables real-time performance; customizable buttons on the screen; preset chords, scales, solo phrases, and songs; fretboard can be moved by dragging, side scroll indicator, or tilting the device; (iOS 7.0 or later).

My Bass (http://trajkovski.net/) Virtual bass guitar simulation for Android; play bass guitar on device like on a real bass guitar; four bass guitar types include electric fingered, acoustic bass, electric picked, and electric slapped; two play modes include solo mode and tapping mode; studio-quality sound; multi-touch support; six sound effects; play velocity; string muting; string damping; string bending; string animation; right- and left-handed; five neck skins; four, five, or six frets; capo option; virtual vibrato arm and volume knob; tablet/pad support; hardware acceleration; advanced haptic feedback; (Android 2.1 and up).

GUITAR APPS

Classical Guitar (http://apps.h2indie .com) Play a virtual classical guitar on Android phone or tablet; nylon strings; responsive and easy to play; works best with big-screen devices and tablets; includes classical guitar, acoustic guitar (full version), electric guitar (full version); realistic strumming actions and string vibration; configurable chord sets; two different play modes: top and side; support for left-handers; capo (full version); (Android 2.2 and up).

Electric Guitar! (http://onbeatapps .com/) Play the electric guitar on the iPhone, iPad, and iPod Touch; sounds are recorded in a professional recording studio; (iOS 5.0 or later).

Guitar Room Pro (http://keynotestar.com/?portfolio=guitar-video) Form or select chords on real guitar images with studio-recorded sounds; virtual guitars collection to fill room wall with guitars for different music styles; high-definition, professionally developed images; play studio-recorded guitar sounds note by note; choose chords or form chords and app will match finger positions with smart chords feature; customize screen with up to thirty chord buttons or minimalist text; learn and practice dozens of popular song chord sets and progressions with hundreds of chords; can edit progressions or add own chords; share songs with users or learn to play uploaded songs by watching how they are played; record and share with others on Facebook and Twitter by uploading recording to SoundCloud directly from the app; (iOS 6.0 or later).

Guitar World Magazine (http://www.guitarworld.com) Publication for rock guitarists; unique instructional content; exclusive photography; interviews with leading celebrity guitarists; offers lessons that help players refine their craft; in-depth reviews of the latest guitars, amplifiers, and effects; current happenings from around the rock scene; celebrity columnists and one-on-one conversations with artists; (iOS 5.0 or later).

Guitarism (http://www.rhism.com) Pocket acoustic guitar; supports Audio-Copy and General Pasteboard; Audiobus supported input; natural interface with high sound quality; includes fret muting, hammer-ons, pull-offs, and up to eighteen active chords; play along with songs from music library; save chord presets per song; record, share, and watch performances; get feedback and stats on published performances and climb the charts; integrate with DAW via AudioCopy and export to iTunes File Sharing and Apple Garageband; with Smart Strings IAP can finger-pick, play arpeggios, and strum; demo video; with Quadroplay IAP can simultaneously play up to three additional instrument apps; (iOS 6.0 or later).

Guitarist (http://moocowmusic.com/OtherApps/index.html) Innovative guitar simulator; record and save songs for later playback; fretboard scrollable up to the twenty-first fret; display between three to ten frets on-screen; record two guitars, rhythm and lead, each with its own effects, into one song; innovative guitar interfaces; pitch bend by dragging strings; tilt for wahwah and whammy effects; choice of acoustic and electric patches available; experienced players can use the app on the road to experiment with chord progressions and record riffs or solos on a musical notepad; novices can program in existing tablature and chords to popular songs and then give performances to their friends; features a selection of guitar interfaces that have been especially tailored to the iPhone; Manual Fret Guitar is an emulation of a real guitar; fret and strum chords or slide solo notes; Hammer-On Guitar assists playing by allowing user to concentrate on fretting notes; press on a fret to instantly play the note; allows complex solos to be played at speed; the full twenty-one frets and open string are available; Tab Guitar removes the concept of fretting strings, allowing the user to concentrate on plucking notes; songs are pre-programmed into the Tab Guitar and then played back live by tapping or strumming at the pickup; uses standard guitar notation tablature; find tablature to almost any song on the Internet and use the Tab Guitar to play songs live; Tab Guitar can be used as a songwriting tool with eighteen of the most common chord types across all keys accessible with a finger-press; can program in own chord pattern via a representation of the fretboard; program in complex chord progressions and then strum them back live at full speed; Scale Guitar allows the user to play runs through a large number of musical scales in any key; both the key and scale can be changed at any time and the scale can be played by running a finger up and down the guitar neck; Scale Guitar can be used as a live performance instrument as well as a useful teaching aid by displaying the notes in each scale and key; does not support tuning for left-handed playing; (iOS 3.0 or later).

GuitarStudio (http://frontierdesign.com/Guitar/) Learn and play favorite songs; acoustic guitar app with virtual fretboard; optimized for performance; perform songs with any combination of chords and melodies on easy-to-use interface; six acoustic guitar sounds, including six- and twelve-string steel and nylon guitars; notes never drop out due to limited sample length; adjust tuning to match recordings and other instruments; reverb effect simulates room ambience; create and save own custom chords; strumming position controls loudness of chords; chords are reordered for fast access; includes twenty-one scales, available in twelve keys; play melodies; hammer-ons, pull-offs, and slides; user-selectable string labels; slide finger between buttons for fast chord transitions; shortcuts for fast song editing and button reordering; support for the left-handed guitarist; record performances with optional metronome; play along with recordings; speed up or slow down performance after recording; combinations of chords and scales can be saved as songs; back up/restore songs to the web; songs can have multiple pages of buttons; built-in support for SoundCloud; export recordings to computer via iTunes; supports Audio Copy/Paste between other apps; jukebox of shared songs from around the world; share performances and be rated by other guitarists; download songs and performances from other users; check out the AirPlay charts; built-in music player; play along with library or imported tracks; supports AudioCopy and General Pasteboard; (iOS 4.3 or later).

GuitarTone (http://www.sonomawireworks.com/iphone/guitartone/) Audiobus supported input and effects; inspired by iconic and boutique amps and effect pedals; wide variety of high-quality tones; interpretations of essential vintage and custom amps, pedals, and microphones; three amps and three effects with more available for purchase; Universal app; (iOS 4.3 or later).

Gypsy Jazz Guitar Secrets Magazine (http://www.playgypsyjazzguitar.com) Explore the world of gypsy jazz with live performances, cutting-edge lessons, and exclusive music and interviews; free mini-issues each month; digital-only monthly magazine featuring articles and interviews with guitarists from around the world; (iOS 5.0 or later; Android 4.1 and up).

iFretless Guitar (http://ifretless.com) Guitar fretboard redesigned for maximum playability on iOS devices; replaces two-handed playing and strumming with a one-touch-per-note design that enables fast, precise playing of melodies, scales, and arpeggios and real-time control over the pitch of each individual note; Audiobus and Inter-App Audio support; (iOS 7.0 or later).

My Guitar (http://trajkovski.net/) Virtual guitar simulation for Android; play guitar on device like a real guitar; four guitar types include steel acoustic, nylon acoustic, clean electric, and muted electric; four play modes include solo mode, tapping mode, chord mode, and combined mode; 325 chords; studio-quality sound; multi-touch support; six sound effects; play velocity; string muting; string damping; string bending; string animation; right- and left-handed; five neck colors; four, five, or six frets; capo option; virtual vibrato arm; virtual volume knob; chords load/save; SD card support; tablet/pad support; hardware acceleration; advanced haptic feedback provided by Immersion's MOTIV engine; (Android 2.1 and up).

Nylon String Guitar (http://apps.h2indie.com) Play a virtual nylon-string guitar; responsive and easy to play; realistic strumming action and string vibration; configurable chord sets; three different playing modes include top, side, and fretboard; sized proportionally to a real classical guitar; support for left-handers; dynamic touch sensitivity; capo; (iOS 4.0 or later).

Real Guitar (http://realguitarapp.com) Realistic guitar simulator app featuring a user-friendly interface; all the notes have been recorded from a live guitar; strum, pluck, and strike the strings to play the chords of any complexity; figure out or make up tunes; learn and master new chords and jingles; solo mode; right-handed and left-handed; database of chords with tabs; HiFi sound; songbook; two types of guitars include nylon and steel strings; (iOS 5.0 or later).

Real Guitar—Android (http://www.rodrigokolb.com.br/#apps) With acoustic guitar, clean electric guitar, and overdrive/distorted electric guitar sounds; two modes: solo and chords; play live music; custom chords sequence; multi-touch; realistic guitar sounds; studio audio quality; record mode; play in loop;

works with all screen resolutions; for guitarists, musician, performers, and artists; (Android 2.1 and up).

Six Strings (http://apps.h2indie.com/strings) Play guitar and other string
instruments on iPad; instruments included are acoustic guitar, classical guitar,
electric guitar, mandolin, banjo, ukulele, steel drum/pan, and drum pad; swipe
fingers across the strings to strum or pluck like a real string instrument; for
jamming, live performances, and assisting in learning how to play a real guitar
or other string instruments; realistic strumming actions; realistic string vibration; swipe fingers across the strings to strum or pluck; record and save loops;
configurable chord sets; three different playing modes for stringed instruments;
instruments are sized proportionally to the real ones except for the steel drum;
metronome; support for left-handers; real-time digital effects including distortion, chorus, flange, tremolo, and echo; twenty-three reverb presets; share
with other users; capo for electric, classical, and acoustic guitars; video demo;
(iOS 6.0 or later).

Strum Stage (http://www.strumstage.net) Sampled guitar sounds and simple
controls; feels like playing a real guitar; grip device like a real guitar and strum
fingers across the screen; supports AudioCopy and General Pasteboard; (iOS
4.3 or later).

Tiny Guitar (http://www.squarepoet.com/tinyguitar/) Tap or strum anywhere to play the notes; play fast or slow; play favorite songs; more than 170
famous tunes and riffs; (iOS 5.0 or later).

Virtual Electric Guitar (https://sites.google.com/site/appsmz/virtual-electric-
guitar) Virtual electric guitar for Android; for beginners and professional players; realistic sound; large chord library; multi-touch; sound effects; can be used
as a guitar tuner; two modes include solo and chords; (Android 2.2 and up).

Virtual Guitar (https://sites.google.com/site/appsmz/virtual-guitar) Turns
Android phone into a classical guitar; for beginners and professional players; realistic sound; best with headphones or external speakers; large chord
library; multi-touch; can be used as a guitar tuner app; two modes: (1) solo
mode: touch string on selected fret to get sound; can go to settings and reduce
number of frets displayed on the screen; (2) chords mode: select the chords
needed and play; chords can be played automatically after selected; (Android
2.2 and up).

WI Guitar (http://www.wallanderinstruments.com/?mode=apps) App is a
virtual acoustic guitar; Core MIDI guitar controller over WiFi; Apple Core
MIDI compliant; (iOS 3.0 or later).

GUITAR LESSONS APPS

Beginner Guitar Songs (http://www.guitarjamz.com/app/beginner_guitar_ songs/#contact) Learn to play the guitar; beginner video guitar lessons one through eight; seven most important guitar chords; six strumming and rhythm patterns; learn to play nine easy guitar songs; beginner manual includes chord charts and important guitar info; offline lesson library; optimized video stream- ing; access to more than nine hundred lessons covering blues, rock, jazz, and more; (iOS 6.1 or later).

eMedia Guitar Method for iPad (http://www.emediamusic.com/guitar- lessons/ipad-lessons-for-beginners-guitar-method.html) More than 120 step-by-step lessons cover basics like stringing a guitar and playing simple chords to techniques such as strumming styles, playing melodies, finger-picking styles, and reading tablature; play songs made famous by artists such as the Rolling Stones, Bob Dylan, and Guns N' Roses; step-by-step approach; more than fifty full-screen videos from professional guitar instructor Kevin Garry, Ph.D.; introduces each new technique using split screens and close-ups of both hands; practice the technique slowly with exercises, then play a song that includes the new technique; more than sixty songs cover folk, blues, classical, and rock styles; modern rock chapter teaches various strumming styles, movable power chords, and power chord riffs; with every song and exercise, animated fretboard guides user through the fingerings while music tracking shows what to play; touch-screen interface; highlight a difficult section of music by touching it with two fingers and pinching, then loop and repeat it to learn; can control the speed of songs; no previous guitar experience is needed, just an acoustic or electric guitar; portable and convenient guitar instruction; learn anytime, anywhere; (iOS 4 or later; Variable Speed Playback requires iOS 5 or later.)

Fret Tester (http://www.guitargames.net/games/fretTester/index.php) Fretboard trainer that will teach user to read music on the guitar, mandolin, or bass; covers learning the notes on the fretboard and reading standard music notation; written by educator William Wilson and is based on years of music teaching experience; clear and straightforward; buttons are easy to push with no clutter; four instruments: guitar, bass, mandolin, five-string bass; instrument and game statistics to chart progress; fretboard charts for all instruments; custom tunings and left-hand mode; HD graphics; four modes of play include: Name Note: a dot is shown on the neck and user names it by letter name; Find Note: a note name is shown and user picks from four locations on the neck; Notation: a note is shown in standard notation and user must tap the on-screen fretboard in the correct location; Notes on Staff: a treble or bass clef is shown and user must name the corresponding letter; (iOS 3.1 or later).

Gibson Learn & Master (http://www.learnandmaster.com/guitar/) Essential tools for becoming a better guitar player; full-length lessons; StudioShare allows the novice or professional musician to record, mix, and share music with friends, fans, and fellow musicians anywhere in the world; global music collaboration; record music using built-in microphones, external microphones, or with add-ons such as the iRig and AmpKit; search system to find other users and music to listen and contribute to; messaging system to make introductions, swap ideas, and contact each other through e-mail; all music files are stored on StudioShare's secure online servers, accessible at any time; create, upload, or delete files from the StudioShare app or from a web browser; paid accounts have a higher-quality recording format; professional-quality chromatic tuner allows user to tune any instrument using an iOS device with a built-in microphone; alternate tuning settings to tune guitar to Open G, Low C, Drop D, E♭, and others; standard mode tuner for tuning each string to a sample tone; metronome with multiple time signatures, visual and audio options, and a tempo tap pad; printable resources to accompany lessons; (iOS 4.0 or later).

Guitar 101—Learn to Play the Guitar (Pacific Spirit Media) YouTube videos of guitar lessons for beginners; requires an Internet connection; (iOS 4.0 or later).

Guitar Interval Ear Trainer (https://sites.google.com/site/guitarinterval/) One of the first steps in ear training is recognizing intervals, which means knowing what the distance is between two notes; because every melody, chord, or scale consists of a series of intervals, this is a very useful and fundamental skill in music; app helps user to identify the intervals and learn how to find them on the guitar; knowledge of intervals on the guitar helps to figure out chord progressions more easily; reproduce melodies and get a better understanding of written music; seventeen selectable intervals, from unison, minor second through major tenth; choose between melodic and harmonic intervals, ascending and descending; graphical representation of intervals on the guitar; chord root note can be chosen over two octaves; automatic or locked portrait and landscape mode; optional voice function; practice and test modes; intervals are grouped by difficulty level; each group has six progress indicators corresponding to the different interval types including melodic ascending/descending, harmonic ascending/descending, random melodic, random harmonic/melodic; optional reference melodies for each interval can be played; in practicing harmonic intervals, the interval is classified as consonant/dissonant; more than ten alternate tunings including the NST (new standard tuning) and all fourths tuning; option to play only tapped notes; useful in figuring out melodies or trying alternate tunings; in practice mode the root can also be played as the second note in the interval; to be in line with written music, enharmonic equivalent notes are used to have classical interval names; intervals can be practiced and tested in a scale context; to strengthen the scale interval knowledge, the partial scale runs of the interval can be played automatically backward and/or forward for a chosen scale; more than fifteen of the most important scales are implemented and can be visualized on the fretboard; scale notes are clustered and colored to show clearly the root note for which the interval applies; support for the right-hand player, the left-hand player with a right-handed guitar, and the left-handed player with a left-handed guitar; use the left-handed option in settings and turn the device by 180 degrees for guitar orientation; Universal app; (iOS 5.1 or later).

Guitar Lesson Apps by TrueFire (http://truefire.com/apps/) Instructional guitar lesson apps for iPad, iPhone, and iPod touch; all apps include video guitar lessons and supplementary learning materials; learn guitar on the go; apps include: Lick of the Day, 50 Blues Rhythms, 50 R&B Guitar Licks, Robben Ford's Blues Revolution, Learn Guitar in 21 Days, 50 Progressive Blues Licks, 50 Texas Blues Guitar Licks, Guitar Player's Top 40 Essential Guitar Licks, 50 Blues Guitar Licks, 50 Slow Blues Guitar Licks, 50 Jazz Guitar Licks, 50 Shred Guitar Licks, 50 Rock Guitar Licks, 50 Funk Guitar Licks, 50 Metal Guitar Licks, 50 Acoustic Guitar Licks, 50 Blues Rock Guitar Licks, 50 Country Guitar Licks, and JamBox; (iOS, varies by app).

Guitar Lessons Free (http://www.nadstech.com/download-free-apps/music-apps) Learn to play the guitar; complete and easy book of lessons; includes an electronic book that will teach user how to read and write tablatures; contents include: Introduction, Structure of the Guitar I, Structure of the Guitar II, Structure of the Guitar III, How to Hold the Guitar I, How to Hold the Guitar II, Tuning I, Tuning II, Notes and Chords, The Fret board I, The Fret board II, The Fret board III, Tablature. Structure and Scales I, Tablature. Structure and Scales II, Tablature. Structure and Scales III, Tablature. Symbology I, Tablature. Symbology II, The Right Hand I, The Right Hand II, Pick Technique, Chromatic Exercise I, Chromatic Exercise II, Chromatic Exercise III, First Position I, First Position II, The DO Major Scale I, The DO Major Scale II, The First Melody, First Chord, Basic Chords I, Basic Chords II, Chord Harmonic of DO I, Chord Harmonic of DO II, Arpeggios I, Arpeggios II, Bar Chords I, Bar Chords II; Guide to Tab Notation; learn to read and write guitar tabs; (Android 1.6 and up).

Guitar Sight Reading Trainer (https://sites.google.com/site/guitaratsight/home) Helps user to find the notes on every part of the fretboard with a special focus on position playing; practice and test/quiz mode; four modes of use include: (1) locate note on fretboard based on note in staff; (2) locate note on fretboard based on note name and pitch; (3) find name of note on the staff without fretboard; (4) find name of note presented on fretboard; frets are grouped to restrict the number of notes to be practiced and tested; fret group is highlighted as an area on the fretboard; progress indicator of sight-reading skills is maintained per fret group; longest streak record per fret group is saved as a measure for sight-reading accuracy; stats are used in test mode to concentrate on problem notes; notes to be tested can be restricted to only naturals, sharps, flats, or from a selected scale; nearly twenty scales and ten different roots are available; highlighted fretboard area can be shrunk by selecting a subset of the strings or a single string; more than fifteen alternate guitar tunings, e.g., drop D, NST (new standard tuning), and all fourths tuning; support for the right-hand player, the left-hand player with a right-handed guitar, and the left-handed player with a left-handed guitar; use the left-handed option in settings plus turn the device by 180 degrees to get a suitable guitar orientation; option to have the fretboard shown in perspective in landscape mode; open string labels can be hidden, which provides a more realistic test; support for enharmonic equivalent notes, e.g., both D♯ and E♭ are used in practice and test mode; Universal app; (iOS 5.1 or later).

Lick of the Day (http://agilepartners.com/apps/lickoftheday/) Guitar learning experience; more than fifty courses spanning more than 1,200 lessons; backing tracks, full-screen tab, and Retina display support; genres covered include blues, rock, jazz, country, bluegrass, fingerstyle, rockabilly, RnB, and acoustic lessons; each Lick Pack features twelve to twenty-five lick video lessons, about sixty to ninety minutes of video from a skilled instructor, along with matching interactive tab/notation, text narrative, and backing tracks; many Lick Packs are presented in sets of two volumes per genre, where the second volume builds on the concepts and techniques that were presented in the first; supports dual instructional modes to optimize the learning experience; in Instructor Mode, the student follows along with the instructor via video, with synced tab or music notation and a fretboard display showing the notes being played by the instructor; in Practice Mode, the student controls the tempo of interactive tablature playback and also has access to the backing track audio to practice what was shown in the video; powered by the live tablature rendering engine from TabToolkit; renders tablature dynamically, including realistic guitar audio and advanced features such as A/B looping and speed control; (iOS 6.0 or later).

Rock Prodigy (http://rockprodigy.com/get-the-app/) Customized to user's skill level; real-time feedback; learn chords, scales, rhythm, technique, and theory; structured courses organized in a daily calendar; learn from teachers trained at Musicians Institute; lessons and songs are set within a teaching system developed by Dr. Richard Gard of Yale University; checks to see what user already knows; practice skills; focused on the goal of making music; challenges decide when user should skip some lessons; structured exercises develop strong fingers, technique, and quick reflexes; exercises inside of music games to develop hand-eye coordination; teaches important concepts and skills that help users play hundreds of songs; listens to every note; provides immediate feedback; awards points for playing correctly; tracks progress automatically; pitch detection technology; about five minutes each lesson; for guitar input with an acoustic guitar, need an internal or external computer microphone; for electric guitar input, need an adapter like iRig or Apogee Jam or an amp loud enough for the device's microphone to pick up; headphones recommended when using an acoustic guitar; (iOS 5.0 or later).

GUITAR SONGS AND REPERTOIRE APPS

Classical Guitar Music: Master (Jiang Yazhou) Contains one hundred tracks of the best guitar music in the world played by the greatest guitarists; includes CD 1: Baroque and Before, CD 2: Spanish Guitar Vol. 1, CD 3: Spanish Guitar Vol. 2, CD 4: Rodrigo Concertos, CD 5: Latin American Guitar, and CD 6: The Modern Guitar; listen anywhere without the Internet; background playing on both home screen and lock screen; playback modes: loop—shuffle—repeat; for all iOS devices; Retina support; (iOS 4.3 or later).

Guitar Complete (http://www.betterdaywireless.com/BetterDayWireless/BDW.html) Learn more than five hundred songs included for free with the app from children's classics to classical favorites; songs for kids, songs for pros, songs for fun, and songs for the holidays; songs include: "Star Spangled Banner," "Canon in D," "Fur Elise," "Silent Night," "Jingle Bells," "Scarborough Fair," and more than five hundred more; twenty-four-fret guitar with multi-touch; real strumming and real fretting; chords; real guitar damping and vibrato; open hammer; pull off; string tuning; four-eight string support; adjustable metronome; resizable frets and resizable strings; customizable fret markers; customizable fingerboard; left-handed support; optional auto-plucking; full-screen support; navigation bar; quick jumping to frets; quick fret shifting; learn songs; two modes of learning; practice strumming; learn both hand positions; lyrics included; (iOS 3.0 or later).

Guitar Riff Pro (http://www.learntomaster.co.uk) Learn to play favorite riffs on the guitar; more than four hundred riffs and songs to learn from different eras and genres; configurable guitar includes multi-touch, glissando, highlighting, and note names; resizable guitar suitable for all devices and tablets; no advertising; play the guitar by ear; high simulator quality and is quick and responsive; optimized for all devices and tablets; adjust the width or height of the fretboard to suit playing style or device; use a clean guitar sound with professionally recorded, digitized Fender Stratocaster electric guitar sounds for each separate note or plug in the distortion; focus on sections of the song to master licks; adjust the speed and volume of the riff for playing along; highlight notes on or off to play by ear; slide the note; standard note tab naming convention; songs cover all genres: Rock, Indie, 60s, 70s, 80s, 90s, Naughties, Modern, Latin, Classical, Film Themes, TV Themes, and Traditional; (iOS 5.0 or later; Android 2.2. and up).

Guitar! (http://www.smule.com/apps) Brings user the experience of musical collaboration between guitarist and singer; play songs with top vocals from the Sing! Karaoke app; choose a singer from more than fifty top performances; create the foundation for the singer while strumming the chords of the song; add personal flourishes on individual strings, vibrato, or emphasis on particular notes; choose from four guitars including acoustic, electric, rock, or classical;

for advanced players or those who want to jam and sing; freestyle mode; new songs and singers added each week; suggest songs in the app; subscribe for access to the entire songbook; to strum, swipe finger across the strings; to pick, tap a single string; change the chord by selecting from the colored tabs on the left; create vibrato by shaking the phone; emphasize a strum by swiping faster and playing louder; (iOS 6.0 or later).

PlayAlong Classical Guitar (http://www.atplaymusic.com/our-apps/playalong-classical-guitar/) Listens to playing and guides user through the selected song; choose the settings that complement skills; cumulative statistics of the total correct notes, consecutive correct notes, and the songs played correctly are available; beginners can choose to have finger charts displayed under each note on the staff as they play; touching the finger chart lets the user hear that note; pushing the play button allows the user to hear the whole song; advanced users can choose to display only the note names under the staff or to see a full screen of music; in Manual mode the user can play the song with their own interpretation, using the display-like traditional sheet music; Stats display presents a list of all the songs, showing which have been played and how many of the notes the user was able to play; History feature keeps track of scores for past performances for each song played; Transposition feature lets the user shift the pitch of a song's notes up or down to customize the range of the song; in Game Center stats are listed on the leader boards and songs are presented as accomplishments; includes some free songs and a chromatic scale; song list includes ratings of easy, intermediate, and advanced; share by e-mailing performances; (iOS 5.1 or later).

The Best Guitar Songs Free (http://www.nadstech.com/download-free-apps/music-apps) Includes lyrics and tablatures for some of the best songs in recent history as well as comments and recommendations; picking, solos, chords, licks, and other useful information; YouTube videos; includes an electronic book that teaches user how to read and write tablatures; does not need an Internet connection; includes 225 songs and a manual to learn to read any type of tab; paid version with no ads; (Android 1.6 and up).

GUITAR TABS, CHARTS, CHORDS, AND SCALES APPS

Achording—Guitar Tabs and Chords (http://php.jamonapps.com/achording/) Meta-search engine of guitar chords and tabs for tabs and chords on some of the best tab sites; all tabs are reformatted to fit in device's screen in both portrait and landscape orientations; auto-scroll feature allows user to start playing songs on guitar without having to manually scroll up or down;

save tabs locally to quickly access them; does not host any files and is not responsible for the content on the sites nor the multiple versions of a song that can be found; (iOS 5.0 or later).

All Guitar Chords (http://www.guitaristsreference.com) More than three thousand commonly used guitar chords; triads in any inversion; chord theory and easy-reference diagrams; (iOS 3.0 or later).

ChordBank: Guitar Chords (http://www.chordbank.com/support) Learn and look up thousands of guitar chords; play electric and acoustic guitar on device; includes major, minor, seventh, and more chords; full fingerings; with Reverse ChordFinder user can tap the fretboard and identify any of the chords inside the ChordBank; clear interface; full left-handed support for all fingerings; strum to hear any chord played from audio samples; (iOS 6.0 or later).

ChordFinder (http://chord-c.com/chord-finder-iphone-app/) Find 2,820 guitar chords in standard tuning, each with five different voicings and photos of suggested fingerings; (iOS 4.3 or later).

Chords (http://adamzap.com/iphone/) Guitar chord reference that offers twenty-eight chord types for each note; chord diagrams appear on a large, readable fretboard; (iOS 7.0 or later).

GChord (http://www.asoft.ne.jp/iOS/GChord/) Chord book for guitar; (iOS 5.1 or later).

Guitar Charts Creator (http://www.guitarchartscreator.com) Create custom guitar charts and diagrams; will help user find the name of the chord; create useful chord charts for practice sessions; chords and text together; create any chord; for multiple instruments including guitar, bass, mandolin, banjo, and ukulele; alternate and custom tunings; AirPrint or e-mail as PDF; comes with more than thirty built-in chord charts featuring all of the most commonly used fingerings for most chord types; for any guitarist; (iOS 5.0 or later).

Guitar Chord (Basic) (http://www.amuze-net.com/en/app/guitarchord .html) Book of guitar chords; improve guitar practice; twelve keys are included; includes three types of each position; 756 guitar chord diagrams; check the sound by tapping and sliding; (iOS 4.3 or later).

Guitar Chord Pro Platinum (http://www.hopefullyuseful.com) Write and learn music; songwriting tool; unique fretboard slider allows instant chord positions and inversions; responsive strumming interface; record own or use in-built strum patterns for one-touch strumming; see the chord structures and experiment; for any orientation including right-handed, left-handed, facing the guitar, or looking down the guitar; five unique guitar sounds all with adjustable reverb and distortion; chord modifier buttons; access to thousands of chords; make unique or impossible to play chords; record song, then change and edit each chord; transpose song to any key; multiple scales to choose from; full Core MIDI out support with the ability to control other apps; MIDI File, Audio Export, and AudioCopy; HD graphics; (iOS 5.1 or later).

Guitar Chord Progression Songwriter (http://www.hopefullyuseful.com) Guitar songwriting tool for all levels; fretboard slider allows instant chord positions and inversions; responsive strumming interface; record own or use inbuilt strum patterns; see the chord structure and experiment; any orientation including right-handed, left-handed, facing the guitar, and looking down the guitar; full HD Retina graphics; chord modifier buttons; access to thousands of chords; five unique guitar sounds to choose from, all with adjustable reverb and distortion; record song, then change and edit each chord; transpose song to any key; multiple scales to choose from; full Core MIDI out support with the ability to control other apps; MIDI File, Audio Export, and AudioCopy; (iOS 5.1 or later).

Guitar Companion (http://www.infinautgames.com/apps-for-musicians/ guitar-companion/) Learn scales up and down the neck and different ways to play them; song idea generators; create songbooks; jot down interesting chords or scales; ad free; skins and beat packs; alternate tunings; patterns feature includes advanced patterns; all access pass; lists of chords and scales; zoom mode; on the neck, display notes, intervals, and finger positions; metronome and drum beats; manually strum the chords and scales to hear how they sound; (iOS 4.3 or later).

Guitar Pro (http://www.guitar-pro.com/en/index.php) Mobile version of the famous Guitar Pro tab-editing program; allows guitarists to enjoy viewing, playing, and writing tablature from their mobile device; (iOS 5.0 or later; Android 2.2 and up).

Guitar Scales and Jam Pro (http://www.learntomaster.co.uk) Learn scales, modes, and broken chords in any position; listen and play back the scales or view the patterns; learn how to improvise and solo in any key; create own riffs by playing to a click and save; optimized for all devices and tablets; adjust the width or height of the fretboard to suit playing style or device; select a scale, mode, or broken chord from the extensive list; listen and watch the notes being played through highlights and then play back; check answers; start with a few notes and then work up to all the notes in the scale; repeat scales ascending, descending, or both; view scale patterns as a reference; in Guitar Jam Game select the scale, mode, or broken chord in which to solo; will highlight all the right notes within the selected key on the whole fretboard; jam by aiming to play from the highlighted keys; record and save favorite jam sessions and play back riffs; each note at each string/fret combination individually recorded in recording studio for a clean guitar sound; choose between clean guitar sound or plug in the distortion; adjust the speed and volume of the riff for playing along; highly configurable; slide the note; highlight the whole fretboard to show which notes are within a chosen scale; play along to different percussions; scales, keys, modes, and broken chords; select the root of the scale and type of scale/key, mode, broken chord, and position from thousands including all the regular (major, minor, pentatonic, blues, etc.) and rare (Persian, altered, bebop dominant, etc.), all the modes (Aeolian, Dorian, flamenco, etc.), and broken chords (major 7, major 7(#5), minor 6); high-quality sound and graphics; (iOS 5.0 or later; Android 2.2 and up).

Guitar Scales Power (Thomas Gunter) Helps any level guitar player improve their playing; great scale reference; knowledge of scales is essential when improvising, doing chord construction, and writing music; each scale is shown in positions covering the complete fretboard in any key for many of the most commonly used scales; allows user to see the note letter names, intervals, and fingerings of scales; scale notes can be played at adjustable speed and highlighted as played to aid in learning; scale positions correspond to the popular CAGED system; the CAGED system chord shape for the current position of a scale is displayed to assist in learning and remembering scale positions in any key; included scales: pentatonic: major and minor pentatonic, blues, major blues; major scale modes: Ionian, Dorian, Phrygian, Lydian, Mixolydian, Aeolian (natural minor), and Locrian; minor scales: harmonic minor, melodic minor ascending (jazz minor); symmetrical scales: chromatic; scale and note playback; simple user interface; all keys and positions of scales listed above in free version; display options include notes, intervals, or fingering; play scale by strumming across strings; scales playback in both directions; tap on a note to play it; right- or left-handed; supports alternative tunings; zoom in or out to see the complete neck; Universal app; (iOS 5.1 or later).

GuitarToolkit (http://agilepartners.com/apps/guitartoolkit/) Improve chord and rhythm skills; learning content based on interactive chord sheets; comprehensive guitar tool; chromatic tuner; large chord, scale, and arpeggio library; flexible metronome; supports all instruments including six-, seven-, and twelve-string guitar, four-, five-, and six-string bass, banjo, mandolin, and ukulele, each with standard and alternate tunings; all features automatically adapt for the selected instrument and tuning; library of more than two million chords, scales, and arpeggios; strum chords and tap notes to hear how they sound; metronome can run in background while using other app features or other apps; chord finder; Universal app; e-mail chord sheets to bandmates or students; interactive chord playback; create drum patterns for practicing and jamming; custom-build patterns with up to thirty-two tracks and seventy-five different drum sounds; create virtual instruments that look and sound like real-life instruments; (iOS 5.0 or later).

Pocket Jamz Guitar Tabs (http://www.synapticstuff.com/support/guitar-tabs-ios) Learn to play songs; guitar learning and reference tool; download, upload, or purchase fully interactive guitar tabs music with adjustable tempo, multiple-track playback, audio playback, and fingering guides; context-sensitive effects legends show user how to play any effects present in a song; dynamic loop points allow user to repeat difficult sections; song library automatically groups related arrangements and includes YouTube videos showing how to play songs and Wikipedia background on song history; high-quality audio playback; interactive tablatures with selectable tracks; hands-free scrolling playback with adjustable song tempo and metronome; included with each download is a starter pack of songs from composers like Mozart and Beethoven; partnered with top music publishers to offer high-quality, interactive editions of authentic song arrangements including lyrics; more than two hundred thousand songs including classical, pop, jazz and blues, Christian and gospel, classic rock, children's songs, country, metal, reggae, RnB and soul; set loop points anywhere in the song, allowing user to automatically repeat difficult measures; dynamic tempo settings; import compatible songs into the application for storage, access, and playback; upload Guitar Tabs in GuitarPro format or Piano Scores in MusicXML format from desktop using WiFi transfer, web, or e-mail; works with the AirTurn Bluetooth page turning device to offer a wireless solution for turning pages or scrolling music without removing hands from the keys while playing a song; hi-res Retina interface; full-page sheet music rendering; beginner, intermediate, and advanced arrangements for guitarists of all levels; in-song, content-sensitive chords and effects library; chord fretboard for practicing fingering; (iOS 6.0 or later).

Progression (http://www.notionmusic.com/products/progressionipad.html) Guitar tab editor and guitar tab creation tool for any guitarist or bassist; write, edit, and play back guitar tablature or regular notation using real audio samples; capture musical ideas; user-friendly interface and simple, interactive entry fretboard; hear music performed with real audio samples including acoustic guitar, electric guitar, and bass guitar; select to add additional instruments such as classical guitar, acoustic finger-style guitar, mandolin, and banjo; search for tabs online; view some of the most popular tabs; share work with anyone by sending a file they can view, hear, and edit; e-mail an image file to print; enter, edit, and play back tab, notation, or both; open Notion and

Progression files; open Guitar Pro files versions three-five; open MusicXML files; open MIDI files; MIDI step-time entry with MIDI device; record real-time MIDI input into score; use on-screen virtual instruments to hear sounds before entering them into score; clean and intuitive user interface; support for Retina display; interactive twenty-four-fret guitar fretboard; quick and simple selection palette; distortion and reverb effects; full audio mixer; file sharing; export and e-mail WAV, MIDI, MusicXML, and Notion files; full range of guitar functions and articulations including bends, vibrato, slides, hammer-on, pull-off, mutes, whammy bar techniques, bass slap, harmonics, and more; drag and drop cursor option; insert text; quick undo and redo functions; chord and melody modes; delete and erase capabilities; help files; enter and edit title and composer information; in-app purchases to add more instruments and gear; free scales and exercises included; (iOS 5.1 or later).

ScaleBank: Guitar Scales (http://www.chordbank.com/) Major, minor, and pentatonic scales; play guitar better, design new riffs, and practice anywhere; includes acoustic and electric guitars; full support for left-handed players; scroll up and down a twenty-four-fret fretboard; clear audio samples of all sixteen included scales; support for major, minor, melodic minor, melodic major, whole tone, pentatonic major, hexatonic blues, pentatonic minor, Dorian, Phrygian, Lydian, Mixolydian, Locrian, Egyptian, Persian, and Enigmatic scales in all keys; (iOS 7.0 or later).

Songsterr (http://www.songsterr.com/a/wa/app?device=ipad) Access to five hundred thousand guitar, bass, and drum tabs and chords; subscription based; one version of tab per song; legal app where music creators get paid; tab player; realistic guitar engine; learn and play along; multiple instruments; most songs have tabs for individual instruments (guitar, bass, drums, vocal, etc.); Half Speed allows user to slow down tab playback to learn difficult parts; Loop plays back selected measures; Offline Mode allows user to view and play back previously opened tabs offline; cache size can be controlled via Settings app; Solo allows user to hear only the instrument they are learning; Count In gives user time to prepare for playing after tapping Play; Chromatic Tuner tunes guitar via mic or adaptors like iRig; Background Audio Support to play along using iRig and Amplitube; Instrument Filter allows user to filter out songs that don't have tab; access tabs viewed recently; access favorite tabs and synchronize with website; browse tabs by tags (rock, metal, blues, jazz, etc.); Universal app; (iOS 6.0 or later; Android 2.3 and up).

Tab Toolkit (http://agilepartners.com/apps/tabtoolkit/) Universal app; jam along with favorite songs; buy music in the Tab Store, upload from computer, or search online; includes speed control, A/B looping, full-score music notation with all instruments, and real-time instrument guides; select part to learn, slow it down, and loop it with full control of each track's audio; on guitar, bass, drum, and keyboard tracks, instrument guides show where to put fingers; in the case of drums, shows which drums to hit; Tab Store includes hundreds of affordable, officially licensed, accurately transcribed songs; access own guitar tab collection, supporting text, PDF, and multi-track rich tab formats; full scores for favorite songs, with complete notation for each guitar, bass, drum, and keyboard track, as well as vocal tracks with lyrics; multi-track audio synthesis based on high-fidelity audio samples of more than 190 instruments; high playback quality; music notation of high visual quality; full resolution on all Retina displays; supports several multi-track music notation formats, including Guitar Pro, Power Tab, and Tab Store format; supports text tabs and PDF-based sheet music; text and PDF tabs do not support advanced capabilities such as multi-track notation and audio playback; (iOS 4.3 or later).

TEFview (http://www.tabledit.com/ios/tv_e.shtml) View, print, and play tablatures created in TablEdit Tablature Editor on iPad, iPhone, or iPod Touch; supported file formats include TablEdit, ASCII, ABC, MIDI, and Music XML; tablature and/or standard notation display; MIDI playback with full speed control; ability to assign the volume and the MIDI instrument for any track in real time; metronome and count down; embedded help; file manager; MIDI file import; WiFi file sharing; mark the beat for one measure before starting playback; subdirectory management; playlist; PDF export: PDF can be e-mailed, printed to a compatible AirPrint printer, or opened in a third app; open PowerTab files; change background color; (iOS 4.3 or later).

Ultimate Guitar Tabs and Ultimate Guitar Chords (http://app.ultimate-guitar.com/apps/) Mobile version of the world's largest catalog of guitar and ukulele chords, tabs, and lyrics; learn how to play favorite songs on acoustic, electric guitar, or ukulele; more than eight hundred thousand tabs and free daily updates with recent hit songs; includes popular songs from the latest releases; chords, notes, and guitar scores for more than four hundred thousand songs; chord diagrams with note placement on fretboard; each chord features multiple variations; daily updates; offline access to favorite tabs; quick search by type, difficulty, tuning, and rating; auto-scroll feature; extensions include Tab Pro with access to more than 150,000 interactive tabs with realistic sound playback; multiple instrument tracks available for each tab (guitar, bass, drums, etc.); loop and tempo control; accurate tuner; precision metronome; chords library with more than two million chords; with tab tools can print and share tabs; change font style and size; Fit To Screen feature; export tabs and chords to Dropbox and send via e-mail; (iOS 5.0 or later; Android 2.1 and up).

Wild Chords (http://ovelin.com/wildchords/) Makes learning to play the guitar easy, fun, and motivating; iPad microphone is used to listen and analyze the played notes and chords; includes one hundred songs from very simple beginner songs to more advanced songs; guitar tuner; tips, tutorials, and suggestions for beginners; profile to track progress; learn the most popular chords; includes melody and scale exercises for lead guitar playing; game is played with a real guitar (acoustic or electric) and requires no additional equipment to be played; (iOS 5.0 or later).

GUITAR TUNER APPS

Acoustic Guitar Tuner (http://innerfour.com/index.php?option=com_conte nt&view=article&id=2637&catid=108:music&Itemid=574) Quick and easy-to-use guitar tuner; press down on each string to hear its tone and tune guitar accordingly; (iOS 4.3 or later).

Guitar Tuner (Workpail) Guitar tuner app for Android; select a guitar string, then tune guitar string to that note; works like a pitch pipe for guitar; graphical guitar plays the guitar notes for each string; (Android 2.1 and up).

GuitarTuna (http://ovelin.com/guitartuna/) Guitar tuner app; built by guitarists for guitarists; simple and easy to use; mini-game to learn guitar chords with chord diagrams; different alternative tuning sets including drop d, open, custom tuning, ukulele tuning, and others; for electric or acoustic guitars; works with built-in microphone; accurate audio signal processing; clear visual feedback with signal history; Auto mode lets user tune hands-free; supports left-handed tuning; learn guitar chords from chord diagram or from sound; works in noisy conditions with background noise cancellation; (iOS 5.0 or later).

Discussion and Activities: Performing Music

$\mathcal{A}$ performer's interest in and knowledge of musical works, their understanding of their own technical skill, and the context for a performance influence their selection of repertoire. Performers make interpretive decisions based on their understanding of the expressive intent of the composer and the context in which a musical work is to be performed. To express their musical ideas, musicians individually and collaboratively analyze, evaluate, and refine their personal and ensemble performances through persistent practice, trying new ideas, and applying appropriate criteria. Musical performances are judged based on the interpretation of a musical work and the performer's technical accuracy, musicianship, and fluency. The context and how a musical work is presented influence how the audience responds. The previous seven chapters, chapters 5 through 11, list apps used for performing music, including singing and playing various instruments.

TOPICS FOR DISCUSSION

1. How do context and the manner in which a musical work is presented influence the audience response?
2. How do musicians improve the quality of their performance?
3. How do performers interpret musical works?
4. How do performers select repertoire?
5. How does making and refining musical instruments respond to the needs of composers and performers?
6. How does the evolution of the guitar, piano, or any instrument relate to the evolution of the performer? In what ways do you think instruments will change in the future?
7. How does the quality of the craftsmanship relate to the quality of the sound the instrument produces?
8. How does understanding the structure and context of musical works inform performance?
9. The human voice is an instrument. What advantages and disadvantages does the human voice have over other instruments? How has human vocal performance changed over time?
10. The size of the orchestra has continued to expand since the industrial revolution, as more instruments have been added to the mix to attract and entertain a larger listening audience. How do you see the orchestra evolving in the future?
11. What are the five categories of instruments? What are two examples from each category?

12. What was the last musical performance you attended? What was the experience like? What style of music was performed? What was the setting?
13. When is a performance judged ready to present?
14. Why does the sound of one violin differ from the sound of another violin? Why do you think many individuals are willing to pay a large sum of money to purchase a handcrafted instrument?

LEARNING ACTIVITIES

1. Given the starting pitch, sing the Solfège to a five-note scale pattern or a complete scale in major or minor, ascending and/or descending, using one of the singing apps.
2. Practice identifying music intervals, pitch memory, and relative pitch by using one of the ear training apps. Study and complete quizzes including interval, chord, and major and minor scale identification as well as rhythmic and melodic dictation.
3. Review the apps available for an instrument and choose several to download. Use the apps to learn more about the instrument, its repertoire, and how to play it.
4. Compare performances of music repertoire you are studying on YouTube. Which performance is the best example of how the piece should be played and why? Do any of the videos show the score or the lyrics?

Part III

RESPONDING AND CONNECTING TO MUSIC

· *12* ·

Preparing and Evaluating
Music Performances Apps

METRONOME APPS

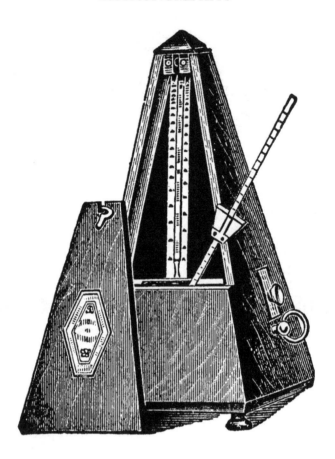

Dr. Bettote TC Metronome App (http://www.ssworkssoftware.com/drbetottetc/ drbetottetc.html) Multi-divisions metronome program; five volume sliders and mute buttons to control quarter, eighth, sixteenth, triplet, and beat note sounds; tap tempo; half-time feel; adjustable swing feel functions; multi-beat mode; can set up to sixteen beats per cycle with the beat sound on any desired beats to create complicated rhythms; all settings, including tempo, can be saved for later recall; alarm timer that syncs with the metronome; quarter note triplet; coach functions; gradual up/down tempo, step up/down tempo, and quiet count; tuning note generator; sync library and settings on multiple devices via iCloud; add own sound via iTunes; (iOS 7.0 or later).

Ludwig Metronome (http://www.ludwig-drums.com/) Full-featured digital metronome application for the twenty-first-century percussionist; (iOS 4.2 or later).

Metronome (http://keakaj.com/metronome/) Allows musicians to mark time by giving a regular tick sound at a selected rate; drag arm horizontally to start and tap to stop; drag arm vertically to change tempo; tap button to access application preferences; (iOS 6.1 or later).

MetroTimer (http://metronomeapp.com) Accurate metronome app; provides highly accurate audio and visual feedback; timer to track practice sessions; (iOS 7.0 or later).

Pro Metronome (http://support.eumlab.com/hc/en-us) Designed for both iPhone and iPad; with time signature setting interface, can customize to any time signature; seven metronome tones are provided; RTP (Real-Time Playback) technology precision; accents of beats can be customized; supports "*f*" and "mute" and "*mf*" and "*p*" indicators; can access subdivision settings including triplet, dot note, etc.; sound feedback; can see and feel the beats; enables Visual, Flash, Vibrate, and Airplay Mode; use Visual/Vibrate Mode when playing loud instruments; Flash Mode can help sync beats; with Airplay Mode can see the beats on projector or Apple TV; can save current parameters including time signature, subdivisions, etc. into a favorites list; can use at any orientation; background playing mode; volume adjustment; tap to calculate bpm; power-saving mode; background-playing mode; Universal app; (iOS 5.0 or later).

Steinway Metronome (http://www.steinway.com/) Full-featured digital metronome for performing artists, music teachers, or music students; dial in the tempo of the piece being practiced or tap along to let the app find it; customize the time signature, visual indicator, and sound options; can change the app's interface; clear, accurate time-keeping; displays the correct musical term for the current tempo; Tap mode lets user find the tempo of a piece by tapping along; works in both portrait and landscape modes; nine time signature settings: 2/2, 2/4, 3/4, 4/4, 5/4, 3/8, 6/8, 9/8, 12/8; eight wood finishes available: African cherry makore, cherry, East Indian rosewood, kewazinga bubinga, macassar ebony, mahogany, santos rosewood, and walnut; interface sounds switch; visual indicator settings: flash on all beats, downbeats, or none; (iOS 4.0 or later).

Tempo—Metronome with Setlists (http://www.frozenape.com/tempo-metronome.html) Accurate metronome; five modes: basic, preset, setlist, practice, and gig; (iOS 5.0 or later).

PRACTICING AND REHEARSAL APPS

Anytune Pro (http://anytune.us/products/anytune-features/) Slow down music, choose the perfect pitch, and learn to play by ear; music practice app for singers, dancers, and musicians of all kinds; learn to play, transcribe, and practice songs by slowing down the tempo, adjusting the pitch, repeating loops, and sharing comments using favorite tracks; plug in and play along with the band using LiveMix and shape the sound with the FineTouch EQ; music practice is perfected; excellent audio quality; import music from Dropbox, e-mail, download from the Web, and more; search music and order playlist; practice at own pace by adjusting the tempo without affecting pitch or sound quality; mark and loop song sections for practicing and sharing; pinpoint instrument in the song visually; record and share performances; ReTune song by adding scripted tempo and pitch adjustments; Step-It-Up Interval Trainer; AirPlay and more; stream live audio directly to other Audiobus compatible apps; transcribe mode to facilitate replay; import audio from videos; export a tuned track via e-mail or "Open in"; LiveMix using an iRigHD, Apogee JAM, or other instrument adaptors; Universal app; (iOS 6.0 or later).

AURALBOOK for ABRSM (http://www.auralbook.com/ad/abrsm/en/pc.php) Helps students prepare for the aural tests of the Associated Board of the Royal Schools of Music (ABRSM) used internationally; analyzes singing performance and areas of improvement on accuracy with original music displayed on the same music sheet; allows students to recognize differences in pitch and beat; instructs in real human voices; detects rhythm and strength of clapping; provides detailed explanations including music history, musical style, instant replay of sections, etc.; selection of more than 1,500 exam questions for all grades; (iOS 4.3 or later; Android 2.3.3 and up).

Decibel Meter (http://future-apps.net/app/158-2/) Uses iPhone's built-in mic to determine the intensity level of sounds it picks up and displays it in dB SPL (Decibel Sound Pressure Level); sensitive enough for professional-level sound metering; (iOS 3.0 or later).

Hit the Note! Learn to Read Music! (http://visionsencoded.com/fun-iphone-apps) Any instrument can be used to play the note, including piano, guitar, or voice, etc.; visual feedback is provided for pitch accuracy; if the red lights are on, go higher or lower as required until the green light illuminates; focus is on letter notation; notation can be set to C D E F G A B or Do Re Mi Fa Sol La Ti; device must have mic input; (iOS 5.0 or later).

MagicSinger (Simple Tools) Turn voice to music; speak to device and app will turn it into music; choose background music according to taste and then tap on the microphone and speak; save the music or setting as a ringtone; (Android 2.2 and up).

Passaggio (http://wivllc.com/products/passaggio/) Tool for lessons, auditions, recording, choir rehearsals, or general practicing; vocal training tool for professionals to use in their teaching; gives the singer a picture of their voice as they are singing; vocal picture is shown against the musical staff so that the singer can tell what is going on and make adjustments on the spot; displays pitch, vocal characteristics, and volume; (iOS 5.0 or later; Mac; PC).

Perfect Vocal (CH Mod, Inc.) Includes pitch shifting, automatic formant correction, and formant shifting; change correction speed; set a recording as a ringtone; automatic pitch correction; re-process recordings to different keys without re-recording; share songs with friends; choose from forty-eight keys; (Android 2.1 and up).

Piano Exam Scales (http://innerwestmusiccollege.com.au) Reference tool to use for piano examination scales; see what is required for different levels and examination boards; prepare for next performance; covers grades preliminary to eighth and exams across the world including AMEB, Trinity, ABRSM, and ANZCA; (iOS 6.0 or later; Android 2.2 and up).

Practice with Sonoptic—Visual Microphone and Music Book (http://sonopticapp .com/) Makes practice sessions more engaging; gives insight into performance; includes more than three hundred exercises, drills, and scales; each one can be customized to user's instrument and level; as user plays, app paints an immediate, artful picture of pitch, rhythm, and dynamics; (iOS 5.1.1 or later).

Sing Perfect Studio (http://www.singperfect.com/) Voice-processing application to make singing sound professional and perfect; "Perfect Tune" processing corrects the timing, melody, vibrato, and loudness of recordings to sound as good as the original singer; adds recording studio effects; build up mixes that sound like records with solos, duets, double tracks, and more; scores the accuracy of each recording and displays the total scores for each mix and each singer; songs can be purchased within the app from a library of hits; variable speed playback control; each mix contains the backing music plus two vocal tracks; can turn the vocal processing on or off for each track; accuracy score shows how much of the melody was sung correctly; (iOS 3.0 or later).

Sing-inTuna (http://www.hotpaw.com/rhn/hotpaw/) Singer can visualize whether or not they are singing on pitch in a scrolling graph; graphs the note being sung on a musical grand staff and color codes pitch accuracy: green for on pitch, blue for sharp, red for flat; (iOS 4.3 or later).

Sing! Vocal Freedom with Professor Angus Godwin (http://www.angusgodwin .com/) Record and send voice for an evaluation and advice; watch critiques of other singers; warm-up exercises; tips on posture; mistakes to avoid; master class; interviews; (iOS 4.3 or later).

Slow Down Music Trainer (http://www.santacruzintegration.com/Music Trainer.html) A music player for musicians, singers, dancers, choreographers, and students; lets listener hear all the nuances; makes it easy to learn new songs by slowing down or speeding up playback without changing the pitch; song will sound the same as the original just faster or slower; use phrase editor to isolate the tricky parts of a song like the solo or a lick and loop that section over and over again; use it to slow down a song to work out a tricky solo, hear the lyrics, or work on a transcription; use it to choreograph a dance routine; makes a great vocal trainer; use the speed up feature to review lecture notes, podcasts, or audio books; works with guitar, bass, trumpet, piano, voice, or any other instrument or audio file; (iOS 4.3 or later).

SmartMusic for iPad (http://www.smartmusic.com/support/) An interactive music learning software; app gives user unlimited access to SmartMusic's extensive repertoire library; full repertoire access requires a SmartMusic account with access to a valid subscription; combines many of the practice features available through a SmartMusic subscription with the convenience of an iPad; open sheet music on screen for more than twenty-two thousand solo and ensemble titles, including repertoire for brass, percussion, strings, woodwinds, and voice; practice with more than fifty method books and nearly fifty thousand skill-building exercises; record and listen to performance; see successes and areas for improvement; if used in an educational setting, students can receive and submit SmartMusic-created assignments as well as custom assignments made with Finale-Created (SMP) files and MP3 audio files; practice more, build skills, and gain confidence; play along with a musical accompaniment while SmartMusic scores performance; (iOS 6.0 or later).

SteadyTune (http://agilepartners.com/apps/steadytune/) Menu bar tuner; accurate, stable, and intuitive; log-ruled tuning strip that animates from out-of-tune dark red to in-tune bright green; tunes all instruments; more than two hundred alternate tunings; adjustable reference pitch; keyboard shortcuts; built-in mic or select an audio interface; level meter shows signal strength; (iOS 10.7 or later).

TUNER APPS

Cleartune Chromatic Tuner (http://www.bitcount.com/support/#Cleartune) Chromatic instrument tuner and pitch pipe; quickly and accurately tune instrument using the built-in mic or an external mic; features a unique "note wheel" interface allowing user to quickly find the pitch, paired with a highly responsive fine-tuning meter for the perfect tune; support for custom temperaments, transposition, notations such as Solfège, adjustable calibration, and more; more power than most pro tuners; simple enough for everyone to use; can tune acoustic or electric guitar, bass, bowed strings, woodwinds, brass, piano, tympani, tablas, and any other instrument that can sustain a tone; built-in temperament for violin family instruments; responsive twenty-five-cent range fine-tuning display; needle damping option; accuracy ±0.01 semitone (±1 cent); selectable temperaments; selectable notations such as Solfège; user-defined temperament and notations; support for transposing instruments; automatic reference note calibration; adjustable A4 calibration in 0.1 Hz increment; pitch pipe/tone generator; selectable tone waveform; automatic or manual note selection; (iOS 4.3 or later).

Epic Chromatic Tuner (http://www.guitarjamz.com/app/epic_chromatic_tuner/#contact) Works with a wide variety of instruments including all guitars, pianos, and acoustic and electric instruments; tune sustained tones to perfect pitch on the fly; (iOS 6.0 or later).

Free Chromatic Tuner: Pano Tuner (http://panoapps.net/index.php) Listens to the sound user makes and shows the pitch; tune instrument accurately by looking at the offset from the desired pitch; quick and sensitive response; wide range of pitch sensor; chromatic tuner; follows any pitch from instruments including guitars, upright basses, trumpets, and piccolos; adjust the Concert A frequency to make instrument in tune with the band; change the sensitivity to tune the instrument more accurately; (iOS 4.3 or later; Android 2.3 and up).

iChromatic Strobe Tuner HD (http://www.hotpaw.com/rhn/hotpaw/) Twelve simultaneous visual strobe tuners for any instrument; fast, accurate, and noise resistant; (iOS 5.1 or later).

insTuner Chromatic Tuner (http://support.eumlab.com/hc/en-us) Chromatic tuner that helps user tune instruments quickly and accurately; tone generator like a pitch pipe with four different wave forms; advanced DSP (Digital Signal Processing) algorithm is highly accurate (±1/1000 semitone or ±0.1 cent precision); supports different tuning modes applicable to different situations, such as instant tuning mode for quick tune and play, fine and strobe modes for accurate tuning, FFT Mode for sound analysis, etc.; "fixed" note wheel helps user find the position of detected pitch easily; can also use it as an electronic pitch pipe to tune by ear, matching the generated tones; suitable for tuning acoustic or electric guitar and bass, ukulele, mandolin, banjo, bouzouki, bowed strings, woodwinds, brass, piano, timpani, and more; can tune any instrument that sustains a tone; supports Line-in Mode, Built-in Microphone Mode, and Clip Microphone; Universal app; supports both landscape and portrait mode; tuning range from C0 to B8, covering the range of almost all musical instruments; Instant Tuning Mode, Strobe Tuning Mode, and Fine Tuning Mode with a highly responsive tuning meter for perfect tuning; FFT Tuning Mode for frequency analysis with log, linear, or note display options; Real-time Spectrogram Mode with log, linear, note, and median frequency display options for real-time audio analysis; adjustable A4 (Concert A) calibration in 0.1 Hz increments; thirty historical temperaments; five notations; needle damping; customizable transposition options for transposing instruments; tone generator with four wave forms; (iOS 5.0 or later).

IStroboSoft Tuner (http://www.istrobosoft.com/) Offers the accuracy of a mechanical strobe tuner; plug in and tune electrics and acoustics with 1/10th cent accuracy; Note/Octave window displays the correct note and octave for the note being tuned; Cents display allows user to see how far out of tune the note is in cent values; tune in auto or manual mode; toggle the display to show: cents, Hz, MIDI note value for the note being tuned; glowing flat/sharp indicators assist when tuning a note far from the target position; noise filter to use with an external mic or clip-on tuning device to help reduce the effect of noise during tuning; input boost raises all input frequencies by +24dB; calibration mode: can be calibrated to an external source guaranteeing 1/10th cent accuracy; drop/capo mode: will auto-transpose notes up or down to one full octave; adjustable Concert A: can change the Concert A reference of the tuner to accommodate tuning to a fixed instrument, such as a piano, or tune instruments that do not utilize the typical Western A440 reference; full-screen mode permits the strobe display to be maximized on screen to allow better viewing from a distance; change strobe display colors to accommodate different lighting environments or set a preferred user color; guitar, bass, brass, and woodwind instrument tuner; not recommended for piano tuning; (iOS 4.0 or later; Android 2.3.3 and up).

Pitch Pipe Now (https://sites.google.com/site/ianjamespiano/blueberry-patch/pitch-pipe-now) Pitch referencing tool in the keys of C and F (chromatic scale); tap the icon to set the pitch for a performance; build chords by choosing any number of note buttons; (iOS 3.0 or later).

Pitch Pro (http://www.frozenape.com/pitchpro.html) Tuner application for the playback of reference pitches for singing and tuning instruments; (iOS 4.3 or later).

ProTuner Lite—Chromatic Tuner (http://www.appsmenow.com/app_page/20176-ProTuner_Lite_-_Chormatic_Tuner) Chromatic instrument

· 13 ·

Listening to Music Apps

CONCERT TICKETS AND TOURS APPS

Live Nation (http://www.livenation.com/mobile) Concerts, festivals, and tickets; (iOS 7.0 or later; Android 4.0 and up).

Master Tour (http://www.eventric.com/) Mobile companion for Master Tour Desktop; tour management and logistics; provides access for crew to itinerary, schedules, travel details, contacts, hotel reservations, maps, and more on device; (iOS 6.0 or later).

INTERNET RADIO APPS

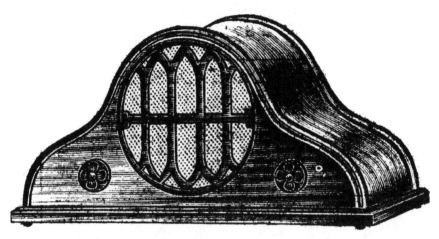

Broadcast—Internet Radio (http://www.domzilla.net/en/iphone-apps/broadcast/) Access more than fifty thousand radio stations for free from the popular SHOUTcast directory; (iOS 5.0 or later).

iHeart Radio—Internet Radio (http://news.iheart.com/go/radio_app/) Popular streaming radio app; (iOS 6.0 or later; Android varies with device).

iTunes Radio (http://www.apple.com/itunes/itunes-radio/) Free streaming radio; (iOS 7 or later).

KandaBi Radio Apps (http://www.kandabisoft.com/) Includes a variety of music genres from around the world; radio stations dynamically loaded; Universal apps; (iOS varies by app).

Live365 Radio (http://www.live365.com/web/components/content/downloads/mobile.live) Browse popular genres; (iOS 4.3 or later; Android varies with device).

NPR Music (http://www.npr.org/music/) Discover new music of all genres; (iOS 5.1 or later).

Pandora (http://www.pandora.com/) Music discovery; search for favorite artists, songs, genres, or composers; create personalized stations; (iOS 6.0 or later; Android varies with device).

Radio Disney (http://music.disney.com/radio-disney) Official Radio Disney app; listen to Radio Disney 24/7; (iOS 6.0 or later; Android 2.2.3 and up).

Radio.com (http://radio.com) Stream CBS Radio stations; (iOS 6.0 or later; Android 2.3 and up).

Rdio (http://www.rdio.com) Social jukebox app includes more than twenty million tracks; (iOS 6.0 or later; Android varies with device).

SiriusXM Internet Radio (http://www.siriusxm.com/ios) App includes sports, talk, comedy, entertainment, and commercial-free music programming; (iOS 4.3 or later; Android 2.2 and up).

Slacker Radio (http://www.slacker.com/) Crafted by hand to deliver the perfect music for any moment; constantly updated; (iOS 6.0 or later; Android varies with device).

TuneIn Radio (http://tunein.com/get-tunein/) Music, news, sports, and talk stations; more than one hundred thousand radio stations; more than four million podcasts; (iOS 7.0 or later; Android varies with device).

MOVIE MUSIC APPS

IMDb (http://www.imdb.com/) Find show times, watch trailers, browse photos, must-see list; collection of movie, TV, and celebrity info; aims to list every detail about every movie and TV show ever made; (iOS 7.0 or later).

Movie Soundtracks (http://www.grupoheron.com) Search a movie in the app and see the songs from the film; extends the musical repertory of films showing almost all the melodies or songs that appear during a movie without limiting to the original soundtrack; (Android 1.6 and up).

MUSIC AND VIDEO STREAMING APPS

Air Media Center (http://www.airmediacenter.com) Multi-platform mobile media center; stream music, videos, and photos from Mac or PC to mobile device; Air Media Server is the server application required in order to use Air Media Center for iOS; Air Media Server automatically transcodes media collection on demand when necessary; (iOS 5.0 or later).

Air Video (http://inmethod.com/airvideo/index.html) Enjoy entire video collection everywhere; solution to watch AVI, DivX, MKV, and other videos; (iOS 4.3 or later).

Amazon Music with Prime Music (http://www.amazon.com/gp/help/customer/display.html?nodeId=200805330&pop-up=1) Listen to or download more than a million songs and hundreds of expert-programmed playlists;

download or stream Amazon Music library from the Cloud or play music on device; (iOS 6.0 or later).

Google Drive (https://www.google.com/drive/?authuser=0) Upload music, photos, videos, documents, and other files that are important; access anywhere on any device; (iOS 7.0 or later).

Google Play Music (https://play.google.com/store/apps/details?id=com .google.android.music) Discover, play, and share music; (Android varies with device).

Music Player (https://play.google.com/store/apps/details?id=media.music .musicplayer) Manage and play music on mobile device; (Android 2.2 and up).

Netflix (https://www.netflix.com/?locale=en-US) Subscription service for watching TV episodes and movies on mobile device; free app as a part of Netflix membership; instantly watch thousands of TV episodes and movies; (iOS 6.0 or later).

Plex (https://plex.tv/features) Organize personal media on any device; stream music, videos, photos, and home movies from home computer running Plex Media Server; (iOS 5.1.1 or later).

Rhapsody (http://www.rhapsody.com/apps-devices/) Listen to more than thirty million songs and albums from favorite artists spanning the decades; (iOS 6.0 or later; Android 4.0 and up).

Songza (http://songza.com) Find favorite artists, playlists, and situations like "driving" and "working out" with a single search; (iOS 6.0 or later; Android varies with device).

Sonos Controller (http://www.sonos.com) System of HiFi wireless speakers and audio components; unites digital music collection in one app controlled from any device; dedicated wireless network; (iOS 6.0 or later; Android 2.1 and up).

SoundCloud (www.SoundCloud.com) Listen to music and audio; discover new and trending music; follow musicians and create playlists; (iOS 7.0 or later; Android varies with device).

Spotify Music (http://www.spotifyartists.com) Search for any song, track, artist, or album and listen; make and share playlists; (iOS 6.0 or later; Android varies with device).

StreamToMe (http://zqueue.com/streamtome/) Play music, video, and photo files streamed over WiFi or Cellular from any Mac or PC running the free ServeToMe server; (iOS 3.1 or later).

VEVO—Watch Music Videos (http://www.vevo.com) App includes a collection of premium official music videos; (iOS 7.0 or later; Android 2.3.3 and up).

WiFi2HiFi 2.0 (http://www.cleverandson.com) Any iOS device can tune in to the music on a Mac; smart streaming technology; plug iPhone or iPad into a docking station, connect it to an audio system, or use its speakers and have Mac's entire music collection, audio streams, and any web radio ready to play; Audiobus supported input; (iOS 5.0 or later).

YouTube (https://www.youtube.com) Upload videos and photos; watch and compare trending videos of singers' performances; subscribe to YouTube channels; watch vintage videos; includes video performances of all music genres; (iOS 6.0 or later; Android varies with device).

MUSIC IDENTIFICATION APPS

Music ID with Lyrics (http://musicid2.com) App instantly recognizes tunes playing on the radio, TV, at the club, or anywhere else; (iOS 5.0 or later).

Shazam (http://www.shazam.com) Identify music and TV shows; will know the name of any song or more about the TV show user is watching; (iOS 6.0 or later; Android varies with device).

SoundHound (http://www.soundhound.com/index.php?action=s.home) App recognizes music playing; tap to identify songs and see lyrics; (iOS 6.1 or later; Android varies with device).

· 14 ·

Music Appreciation, History, and Culture Apps

255

· 14 ·

Music Appreciation, History, and Culture Apps

CLASSICAL MUSIC APPS

Appreciation of Western Music (http://www.ndapk.com) Includes 100 examples of piano music, 112 of violin music, one hundred of guitar, one hundred of cello, and one hundred harmonica songs; tracks are complemented with simplified and traditional Chinese and English introductions; pieces and songs are all composed or performed by masters in music; experience the beauty of Western traditional instrumental music; the piano is one of the most popular instruments in the world and is widely used in classical and jazz music for solo performances, ensembles, chamber music, and accompaniments; the piano is also very popular as an aid to composing and rehearsal; although not portable and often expensive, the piano's versatility and ubiquity have made it one of the world's most familiar musical instruments; the violin is a string instrument, usually with four strings tuned in perfect fifths; it is the smallest, highest-pitched member of the violin family of string instruments, which also includes the viola and cello; the guitar is a string instrument of the chordophone family constructed from wood and strung with either nylon or steel strings; the modern guitar was preceded by the lute, the vihuela, the four-course Renaissance guitar, and the five-course Baroque guitar, all of which contributed to the development of the modern six-string instrument; the cello is a bowed string instrument with four strings tuned in perfect fifths; it is a member of the violin family of musical instruments, which also includes the violin and viola; the harmonica, also called the French harp, blues harp, and mouth organ, is a free reed wind instrument used worldwide in nearly every musical genre, notably in blues, American folk music, jazz, country, and rock and roll; there are many types of harmonicas, including diatonic, chromatic, tremolo, octave, orchestral, and bass versions; a harmonica is played by using the mouth (lips and/or tongue) to direct air into and out of one or more holes along a mouthpiece; behind the holes are chambers containing at least one reed; a harmonica reed is a flat elongate spring typically made of brass or bronze, which is secured at one end over a slot that serves as an airway; when the free end is made to vibrate by the player's air, it alternately blocks and unblocks the airway to produce sound; touch the play/pause button to play the audio; slide the screen to touch leaf through the text; (iOS 3.2 or later).

BBC Music Magazine (Immediate Media Company Limited) The world's best-selling classical music magazine; every single page of *BBC Music Magazine* is included in the app; instantly sample and buy almost two hundred recordings mentioned and recommended in every issue; click on any album cover or on the underlined text and the widget will open up, allowing the user to hear extracts from hundreds of recordings without leaving the app; click on any track to buy it from iTunes, where the choice will be highlighted; for anyone with a passion for classical music; brings the world of classical music to life, including interviews with the greatest artists and features on fascinating subjects as well as the latest news and opinions from around the music world; reviews of more than one hundred recordings, each one rated by the finest writers in the business; for classical music connoisseurs and enthusiasts; (iOS 5.0 or later).

Classic FM (http://www.classicfm.com) Listen to the world's greatest classical music; includes pictures, news stories, and info on every piece played; listen to favorite composers, artists, and shows with live radio stream; listen to live radio on the move in high quality; look at related images, news, and videos for any song currently playing; latest classical music news; share the piece that is currently playing or download it on iTunes; view picture galleries of favorite artists; browse the week's schedule; stream audio to AirPlay device; listen to favorite shows from the past week (iPad only); contains big, full-screen images of artists; uses unique identifiers to improve experience; (iOS 6.0 or later).

Classical Archives (http://www.classicalarchives.com) Large collection of classical music; play/stream works of classical music through queuing player; unique classification system developed by expert musicologists; three modes include: (1) "The Library," where user can select from all the performances of a work in a large collection of recordings organized by composer, category, and composition; includes an immediate list of "The Greats" (composers) and their works, or search for any composer or performer by name; long-tap on one of them to either add it to the player's queue, add it to a playlist, or view and play the entire album it came from; a short tap on a performance sends it immediately to the player or its queue if something is already playing; (2) "Must Know," where the user will be guided to appreciate the most important composers and works of classical music through their latest released recordings; search by composers or by a combination of period including Baroque, Classical, Romantic, etc., and genre including orchestral, chamber, solo, stage, etc.; (3) "Playlists," where user can play suggested playlists or build and play their own; subscribers have unlimited access to play entire tracks, complete works, or albums including operas and multi-CD collections; non-subscribers hear one-minute clips of each track; (iOS 6.0 or later; Android 2.3.3 and up).

Classical KUSC (http://www.kuscinteractive.org) One of the largest and most listened to classical public radio stations in the country; dedicated to making classical music and the arts an important part of people's lives; located in downtown Los Angeles, KUSC has been a broadcast service of the University of Southern California since 1947; (iOS 5.0 or later).

Classical Masterpieces (http://apps.ev-games.com/index.php/cm) Collection of twenty masterpieces of classical music; full biography of all composers with Facebook sharing; listen to music everywhere without an Internet connection; high-quality format; all compositions are performed by well-known musicians and orchestras; more than 150 pieces after upgrade; includes masterpieces of classical music by eighteen of the greatest composers; CD-quality sound; Air-Play, remote control; background playback; playlists; (iOS 5.0 or later).

Classical Masters—Anywhere Artists (http://www.anywhereartist.com/aa/) Collection of relaxing and inspiring melodies featuring free music, news, photos, videos, and more from fourteen of history's greatest classical composers; available for the iPhone, iPad, and iPod Touch; classical music is the art music produced in, or rooted in, the traditions of Western liturgical and secular music, encompassing a broad period from roughly the eleventh century to present times; the central norms of this tradition became codified between 1550 and 1900, which is known as the common practice period, not to be confused with the Classical Era; European music is largely distinguished from many other non-European and popular musical forms by its system of staff notation in use since about the sixteenth century; Western staff notation is used by composers to prescribe to the performer the pitch, speed, meter, individual rhythms, and exact execution of a piece of music; this leaves less room for practices such as improvisation and ad libitum ornamentation frequently heard in non-European art music and popular music; the term "classical music" did not appear until the early nineteenth century, in an attempt to "canonize" the period from Johann Sebastian Bach to Beethoven as a golden age; the earliest reference to "classical music" recorded by the Oxford English Dictionary is from about 1836; given the extremely broad variety of forms, styles, genres, and historical periods generally perceived as being described by the term "classical music," it is difficult to list characteristics that can be attributed to all works of that type; many vague descriptions exist, such as describing classical music as anything that "lasts a long time" or music that has certain instruments like violins, which are also found in other genres; there are characteristics that classical music contains that few or no other genres of music contain; the instruments used in most classical music were largely invented before the mid-nineteenth century, often much earlier, and codified in the eighteenth and nineteenth centuries; they consist of the instruments found in an orchestra, together with a few other solo instruments such as the piano, harpsichord, and organ; the symphony orchestra is the most widely known medium for classical music; the orchestra includes members of the string, woodwind, brass, and percussion families; featured artists include Johann Sebastian Bach, Ludwig van Beethoven, Johannes Brahms, Frédédric Chopin, Claude Debussy, Joseph Haydn, Franz Liszt, Felix Mendelssohn, Wolfgang Amadeus Mozart, Johann

Pachelbel, Maurice Ravel, Robert Schumann, Pyotr Tchaikovsky, and Antonio Vivaldi; (iOS 5.0 or later).

Classical Melody Book (http://themelodybook.com/classical-melody-book/) App includes popular classical sheet music for piano, flute, guitar, violin, voice, orchestra, and much more; download classical scores and view without requiring an Internet connection; play, share, and print; page controls, zoom capabilities, and ability to add comments; (iOS 4.0 or later).

Classical Music Collection (http://wizards.rs/apps/) Radio FM classical music; can be listened to with or without headphones; set a radio alarm; share with friends and family; create a shortcut to favorite live radio stations on home screen; create a list of favorite streaming radio stations by marking with a star; includes the latest popular classical music and radio stations with classical music for kids and classical music for babies; teach children the value of classical music; use app to relax and listen to classical music for meditation; (Android 2.2 and up).

Classical Music I: Master's Collection Vol. 1 (http://www.magicanywhere .com/classical1/) Includes 120 essential classical music masterpieces, professionally handpicked by music experts and performed by famous orchestras around the world; no Internet required; (iOS 7.0 or later).

Classical Music II: Master's Collection Vol. 2 (http://www.magicanywhere .com/classical2/) Includes 110 new and different masterpieces in addition to Vol. 1; (iOS 7.0 or later).

Classical Music Radio (http://developercatworld.blogspot.com) Classical music radio list; more than thirty radio stations with a new station every update; track list; stream recording; add own stations to list; streaming setup for best listening; widget; classical music news feeds; sleep timer; station list; also Game Radio and Anime Radio; (Android 2.1 and up).

Classical Music Radio Petite (http://classical-music-radio-petite.appsios.net/) Listen to the great composers of classical music; selective list with an easy interface; variety of radio stations; stations are dynamically loaded; background music mode; (iOS: 4.0 or later).

Classical Music Shows on Internet Radio (http://www.audibilities.com) Listen to classical music radio shows streaming live on Internet radio; provides up-to-the-minute show schedules; some of the most listened-to shows along with top classical stations such as WCPE and KCSN; suggest shows by e-mail or from within the application; streaming quality depends on network speed; users are responsible for bandwidth use in accordance with contract agreements with their cellular network and any expenses accrued; streaming is provided by commercial and public radio stations and syndicates; linked sites and audio feeds are copyrighted; (iOS 4.2 or later).

Classical Music: Master's Collection, Vol. 1 (http://www.magicanywhere .com/classical1/) Includes 120 essential classical music masterpieces, professionally handpicked by music experts and performed by famous orchestras around the world; no Internet required to listen; children who listen to classical music are proven to be smarter and have higher IQs than the ones who don't, especially when started as babies; from the Baroque to the Modern era including symphonies, intimate chamber works, old-fashioned Baroque music, twentieth-century pieces; relax, energize, concentrate, sleep, study, or work more efficiently; research shows that classical music is the best therapy for tapping into human emotions, helping the brain to stay fresh and focused; search and organize favorite pieces; fall asleep with sleep timer; discover and learn more with YouTube live performances and Wikipedia composer info for each masterpiece in the app; full search capabilities by name and composer; multi-lingual support in ten major languages; music visualizer; non-stop background playing; loop/shuffle/repeat; touch control: double-tap to play/pause, swipe right/left to change soundtrack; supports wireless AirPlay speakers, Apple TV, remote control, and Bluetooth headsets; (iOS 7.0 or later).

Classical Music: Master's Collection, Vol. 2 (http://www.magicanywhere .com/classical2/) Presents one hundred new masterpieces in addition to Vol. 1; professionally handpicked classical music masterpieces; for education and learning; no Internet required; music visualizer; watch each performance live on stage with built-in YouTube Viewer; read composer info and discover more about the masterpiece on Wikipedia; supports speakers, remote control, and Bluetooth headsets; connects into iTunes Store and helps user find different versions of a piece; wide variety of styles and time periods including symphonies, intimate chamber works, old-fashioned Baroque music, twentieth-century pieces, and more; (iOS 4.3 or later).

Classical Music. Listen and Learn. 50 Compositions and Quiz (ADS Software Group, Inc.) Educational collection of fifty famous classical compositions; listen to fifty famous compositions in a retro record player; read short biography of each artist and information about each composition; play "Who is the Composer?" quiz game; share scores; post achievements on Facebook and Twitter and challenge friends; find high-quality versions of compositions on iTunes; includes compositions by Johann Sebastian Bach, Ludwig van Beethoven, Hector Berlioz, Johannes Brahms, Frédéric Chopin, Antonín Dvořák, Sir Edward Elgar, Edvard Grieg, Franz Liszt, Felix Mendelssohn, Wolfgang Amadeus Mozart, Modest Mussorgsky, Johann Pachelbel, Nikolai

Rimsky-Korsakov, Gioaccino Rossini, Camille Saint-Saëns, Johann Strauss, Franz von Suppé, Pyotr Tchaikovsky, Antonio Vivaldi, and Richard Wagner; (iOS 3.0 or later).

Early Music Today (http://www.pocketmags.com/viewmagazine.aspx?catid =1037&category=Music&subcatid=227&subcategory=Classical&title=Early +Music+Today&titleid=345) UK's leading magazine devoted to performers and enthusiasts of early music; key features about the industry; news updates and events listings; reviews of the latest performances, festivals, CD, DVD, and book releases; e-news featuring concert and festival tickets discounts, early-bird bookings with key music venues; entry into competitions; subscriptions; (Android 4.0 and up).

ILoveClassicalMusic (http://www.mortadelanetwork.com) Listen to classical and piano music for free, via streaming, from all over the world; radios from Germany, France, Romania, Argentina, Brazil, and more; (iOS 6.0 or later).

Masterpieces of Classical Music (http://apps.ev-games.com/index.php/cm) Collection of compositions by the best composers of the world; encyclopedia that tells about their lives and work; includes the best works of Bach, Beethoven, Mozart, Handel, Chopin, and many other famous composers; can listen to classical music without an Internet connection; two hundred masterpieces of classical music by fifty-five composers; more than twelve hours of music; great sound quality; division of compositions by composers; detailed biographies; AirPlay; background playback; remote control and playlists support; (iOS 5.0 or later).

Naxos Music Library (http://www.naxosmusiclibrary.com/home.asp?rurl= %2Fdefault%2Easp) Includes more than ninety-nine thousand CDs of music; one thousand CDs added every month; remote streaming of Naxos Music Library (NML) playlists anytime, anywhere with device; download app and try service for free or log in to existing NML account to stream playlists; personal and account playlists are available for streaming through the NML app; use favorite social network to keep track of all the updates; (iOS 3.2.2 or later; Android 2.1 and up).

SyncScore (http://zininworks.com/syncscore/) Provides a wide variety of classical music with its synchronized scores; browse library of handpicked tracks, composed by Bach, Mozart, Beethoven, Chopin, and more; hub for various SyncScore apps that contain full tracks of the album; view full album description and purchase up to nine hours of masterpieces; scroll automatically as the music proceeds; synchronized horizontal progress indicator; intuitive navigation with pagination; multitasking support; background music playback; (iOS 6.0 or later).

CLASSICAL MUSIC APPS FOR CHILDREN

123 Kids Fun Music (http://www.123kidsfun.com/music.html) Music game for toddlers and preschoolers ages one to four; twenty-four instrument sounds including xylophone, drums, guitars, trumpets, flute, bells, and more; fun and easy way to use virtual instruments; tap to hear the sounds; bright, colorful, child-friendly design; (iOS 5.0 or later; Android 2.2 and up).

Classical Kids (http://www.childrensgroup.com/product.php?mode =cat&cid=classic_eng) Producer Susan Hammond's award-winning series; recordings feature a unique combination of music, history, and dramatic storytelling; presents classical composers as heroes for children; all the facts about the period are true; a fictional child leads young listeners into the composer's world; includes quiz questions for each story, resources to learn about the most popular classical composers, a metronome tool to assist with music practice, and an audio recorder to record performances; can purchase any of the seven individual Classical Kids forty-five-minute illustrated musical stories along with read-along scripts including *Beethoven Lives Upstairs, Mr. Bach Comes to Call, Vivaldi's Ring of Mystery, Tchaikovsky Discovers America, Mozart's Magic Fantasy, Mozart's Magnificent Voyage,* and *Hallelujah Handel;* (iOS 4.3 or later).

Classical Music for Babies (Elite Solutions LTD) Discover the transformational power of music for health, education, and well-being; experience "the Mozart Effect" on baby; increase cognitive skills and boost IQ; the Mozart Effect is a term coined by Alfred A. Tomatis for the increase in brain development that occurs in children under age three when they listen to the music of Mozart; the idea for the Mozart effect originated in 1993 at the University of California, Irvine, with physicist Gordon Shaw and Frances Rauscher, a former concert cellist and an expert on cognitive development; they studied the effects on a few dozen college students of listening to the first ten minutes of the Mozart Sonata for Two Pianos in D Major (K.448) and found a temporary enhancement of spatial-temporal reasoning as measured by the Stanford-Binet IQ test; improves test scores, cuts learning time, calms hyperactive children and adults, reduces errors, improves creativity and clarity, heals the body faster, integrates both sides of the brain for more efficient learning, raises IQ scores nine points; (iOS 4.0 or later).

Classical Music for Kids (Inbal tal) Animals play great musical creations including *Swan Lake, Nutcracker, Lullaby;* introduce children to classical music; will teach child the basic information about the musical instruments of the orchestra including how they look, how they sound, what they are made of, and how to play them; introduce children to interesting information about the great composers including a basic biography of the greatest composers' lives, where they came from, how they grew up, and their most famous pieces; it is commonly agreed that classical music has a profound effect on mood, spatial intelligence, memory, and language; listening to music can have positive effects on health; music manipulates moods, enriches story lines, and envelops minds; impact occurs on both babies and adults; (Android 2.2.3 and up).

Classical Music for Mommies (http://www.crazyteabag.com) Classical music has been proven to be the music of intelligence; recent studies have shown that listening to classical music can have lasting effects on the brain; it can stimulate brain function in developing infants and improve comprehension of math and other computing skills in adults; it trains the brain to acknowledge all the parts that work together to make the whole; listening to classical music during pregnancy may reduce the stress, anxiety, and depression that many pregnant women experience; improves comprehension of exact sciences and many others; effects are most pronounced when submerged completely into the music; (Android varies with device).

Kids Musical Instrument Jigsaw Puzzles (http://espacepublishing.com) Complete jigsaw puzzles while learning about musical instruments and identifying the sounds they make; (iOS 5.0 or later; Android 2.1 and up).

Little Beethoven App (http://www.naxos.com/catalogue/apps/app.asp?ID=NAPP0204) Includes eighteen tracks; clear, lively text outline of what the music means; fully narrated; animated illustrations on each screen suited to each piece of music; bookmark favorite tracks; see which tracks have been listened to and which haven't with the progress marker; music includes Symphony No. 5, "Für Elise," *Fidelio, Moonlight* Sonata, *Spring* Sonata, *Pastoral* Symphony, Violin Concerto, and many more; fun and friendly introduction to Beethoven for young children with many famous and well-loved pieces; compact with animated illustrations; ideal introduction for young children to classical music; selection of tracks is carefully tailored for younger listeners, including famous pieces as well as unexpected gems; narrated text is a friendly outline of what the music means, concentrating on the stories behind it, the sounds themselves, and on the composers; simple interface to stop and start tracks and animation or hear the narrated text; children and parents can discover together; shows how some of the music Beethoven wrote is strong and serious and some of it is very gentle; listen to orchestras, pianos, violins, and other instruments performing; (iOS 4.3 or later; Android 2.3 and up).

Little Classical Music App (http://www.naxos.com/catalogue/apps/app.asp?ID=NAPP0201A) Includes twenty-five tracks; clear and lively outline of what the music means; fully narrated; animated illustrations on each screen suited to each piece of music; bookmark favorite tracks; find which tracks have been listened to and which haven't with the progress marker; music includes Vivaldi: *The Four Seasons*, Beethoven: Symphony No. 5, Grieg: *Peer Gynt*, Saint-Saëns: *The Carnival of the Animals*, Dukas: *The Sorcerer's Apprentice*, Prokofiev: *Peter and the Wolf*, Mozart: *The Magic Flute*, Rossini: *William Tell*, and more; fun introduction to classical music for young children with many well-loved pieces; (iOS 4.3 or later; Android 2.3 and up).

Little   Mozart   App   (http://www.naxos.com/catalogue/apps/app .asp?ID=NAPP0202A) Includes seventeen tracks; clear and lively outline of what the music means; fully narrated; animated illustrations on each screen suited to each piece of music; bookmark favorite tracks; progress marker; music includes *The Marriage of Figaro*, Symphony No. 40, *Eine Kleine Nachtmusik*, *Turkish* Rondo, Clarinet Quintet, *The Magic Flute*, *A Musical Joke*, and many more; introduction to Mozart for children; (iOS 4.3 or later; Android 2.3 and up).

Little   Tchaikovsky   App   (http://www.naxos.com/catalogue/apps/app .asp?ID=NAPP0207A) Includes twenty-three tracks; clear, lively outline of what the music means; fully narrated; animated illustrations on each screen suited to each piece of music; bookmark favorites; progress marker; music includes *The Nutcracker*, Piano Concerto No. 1, *Swan Lake*, Symphony No. 5, *The Seasons*, *The Sleeping Beauty*, and many more; fun introduction to Tchaikovsky for young children with many well-loved pieces; (iOS 6.1 or later; Android 2.3 and up).

Magic Flute for Little Composers (http://littlecomposers.com) Custom tailored to make the beginning stages easy and memorable; colorful play pads; built-in lessons make learning easy for everyone; real songs to learn; ear training to build up musical talent; jam along sessions for advanced students; jam module for right- and left-handed children; (iOS 3.2 or later).

Music Instruments for Kids (http://www.theappguy.co/toddler-sounds) App enables toddlers to associate the sounds of common musical instruments; scroll through pictures of a variety of musical instruments and press Play on each to hear the sound; (iOS 5.0 or later).

Music Sparkles (http://www.tabtale.com/app/music-sparkles-2/) Brings a world of music to fingertips from piano and xylophone to electric guitar; tap and play; sparkles are created as user plays; large collection of musical instruments to develop musical hearing; includes fourteen musical instruments: drum set, xylophone with four octaves, classic guitar with twenty-four notes, electric guitar with twenty-four notes, harp with twelve strings, saxophone with three octaves, recorder flute with two octaves, grand piano with four octaves, accordion with two octaves, harmonica with two octaves, pan flute with two octaves, two violins with eight strings, classic pipe organ, synthesizer with four octaves, five music loops; five different music looks including classic guitar loop, authentic African drums, Heep banjo, majestic grand piano chords, and cool drum beat; (iOS 5.0 or later).

Musical Instrument Flashcards (http://babykidszone.com/babygames/bthmusic/index.html) App includes the sounds of sixty different musical instruments with descriptions; five different categories; slideshow feature; (iOS 4.3 or later).

My First Classical Music App HD (http://www.naxos.com/catalogue/apps/app.asp?ID=NAPP0101A) Introduction to classical music for children aged five and above; find out where you hear music, who writes it, and what all the instruments sound like; meet the great composers and learn about the orchestra; tap any words or pictures and hear the text narrated, extracts of music, and dozens of animations and sound effects; animals can dance, sing, and play; children can enjoy alone or together with a parent; includes many different pieces of music; full album of music includes Grieg: *Peer Gynt*, Mozart: *Magic Flute*, Williams: *Harry Potter and the Sorcerer's Stone*, Saint-Saëns: *Carnival of the Animals*, Holst: *Planets*, Stravinsky: *Petrushka*, Prokofiev: *Peter and the Wolf*, and many more; interactive iPad version of *My First Classical Music Book*, published by Naxos; includes three main sections: When? Where?, People, and Instruments; friendly, engaging narration; Questions and Listen For tips from the Music Bird; more than forty different pieces of music featured, in excellent recordings; every instrument in the orchestra is demonstrated; lively character illustrations throughout; (iOS 4.3 or later; Android 2.2 and up).

The Carnival of the Animals App (http://www.naxos.com/catalogue/apps/app.asp?ID=NAPP0601A) Popular musical work by Camille Saint-Saëns enhanced with words and animations; entertaining app; excellent recording of the music; presents brand new verses and animated illustrations; tap the pictures for surprises; (iOS 5.0 or later; Android 2.2 and up).

Toddler Classical Music Jukebox (Tipitap, Inc.) App is an introduction to classical music for young kids; fun and colorful interface; twelve favorites from Beethoven and Mozart to Chopin and Brahms; simple piano renditions for young ears; (iOS 4.0 or later).

Zoola Children's Classics—Classical Music for Kids (http://www.zoolaapps.com/zoola-childrens-classics-2/) Introduces children to classical music with animals playing great musical creations; teaches children the basic information about the musical instruments of the orchestra including how they look, how they sound, what they are made of, and how to play them; introduces children to interesting information about the great composers; basic biographies of the greatest composers lives, where they came from, how they grew up, and their most famous pieces; classical music has a profound effect on mood, spatial intelligence, memory, and language; listening to music can have positive effects on health; music manipulates moods and enriches story lines; languages include English, French, German, Russian, and Spanish; (iOS 5.0 or later).

CLASSICAL MUSIC COMPOSERS APPS

Bach Machine (http://intelligentgad gets.us/apps.shtml#2) Listen to Bach; multi-timbral synthesizer and sound map; orchestrate with fingers; change instruments in real time; playlist contains hundreds of Bach concerti, sonatas, inventions, and more; demo on YouTube; (iOS 5.1 or later).

Beethoven Symphonies (http://dre ampocket.miniban.cn) App includes all nine symphonies; Ludwig van Beethoven (1770–1827) was a German composer and pianist; a crucial figure in the transition between the Classical and Romantic eras in Western art music, he remains one of the most famous and influential of all composers; acknowledged as one of the giants of classical music, he occasionally is referred to as one of the "three Bs" along with Bach and Brahms; he was a pivotal figure in the transition from eighteenth-century musical classicism to nineteenth-century romanticism; his influence on subsequent generations was profound; (iOS 4.3 or later).

Beethoven's Ninth Symphony (http://www.touchpress.com/#our-apps)

App presents four of Deutsche Grammophon's legendary recordings of the iconic work, with the ability to switch between each performance at any point in the music; includes a synchronized musical score guided by expert commentary; graphical BeatMap of the orchestra, highlighting every note; free version of the app includes two minutes from the second movement of the symphony, with all features enabled; the full experience can be unlocked through in-app purchase; no further download is required; four legendary performances of Beethoven's *Ninth Symphony* including Ferenc Fricsay's first stereo recording of the work from 1958 with the Berliner Philharmoniker; Herbert von Karajan's famous 1962 recording with the same orchestra; maestro Leonard Bernstein's video recording from 1979 with the Wiener Philharmoniker; and the 1992 recording on period instruments conducted by John Eliot Gardiner; typeset score, simplified score, and hypnotic BeatMap are synchronized to all performances; specially commissioned synchronized commentary by David Owen Norris; (iOS 6.0 or later).

Binaural Beethoven Pro (http://3hstudios.webnode.com/products/binaural-beethoven-pro/) Relax with Beethoven's sonatas and binaural beats; includes harmonies with natural sounds and relaxing waves; an inspirational tool; makes sixty original sound combinations; includes three impressive sonatas carefully resounded by Pro Tools studio software to produce more colorful sounds but with respect to the original moods and dynamics; three sections provide independent sound sources that can be mixed together and saved as presets with related volume levels; first section includes three sonatas; second section provides binaural beats for Sleeping, Learning Aid, and Meditation; third section includes natural sounds; one hour long; (iOS 7.0 or later).

Guess Composer Classical Music Quiz (http://guess-composer.ru) Classical music is used in films and ads; most people can sing the tunes, but they don't know the composer or the title; app game will help listener to know the most famous works of the greatest composers; one hundred masterpieces of the classics; more than forty classical music remixes; interesting facts about the works; check game results; awards and statistics; for connoisseurs of the classics and for people who have only started to discover them; game rules are simple: listen to the tune and guess its name and composer by choosing one of the four answers; after passing five tunes, listener can analyze mistakes in the results window; (iOS 4.3 or later; Android 2.2 and up).

iTunes U (https://www.apple.com/itunes/) App gives access to complete courses from leading universities and other schools; large digital catalog of free education content; (iOS 7 or later).

Masterpieces of Great Composers (http://www.ndapk.com) Includes hundreds of compositions by Bach, Mozart, Beethoven, and Chopin; the selected pieces are representative works that are often performed; includes brief introductions helping readers to analyze and appreciate the music; on-demand download function for users to select tracks; (iOS 4.3 or later).

Mozart Interactive (http://melodystreet.com/interactive/mozart/) Interactive musical experience where user decides what happens next; includes Melody Street's host Ethan Bortnick and stringed pal Val Violin playing a duet of Mozart's *Rondo Alla Turca*; user chooses between Heidi Horn, Timmy Trumpet, and Febe Flute; orchestrate own version and share with family and friends; (iOS 4.0 or later).

Radio Mozart (http://www.radiomozart.net/Pages/default.aspx) The concept of Radio Mozart is original, unique, and very simple: Mozart and nothing but Mozart, all the time; includes all of Mozart's works, combining all genres: symphonies, concertos, serenades, religious music, and opera; hear the most famous compositions of Mozart; launched in 2010, Radio Mozart is available for free on the Internet; station gets more than forty thousand hits per day, for an average listening time of forty-five minutes; one of the most listened-to classical radio stations; a community radio station based in Marseille, France; (iOS 6.0 or later).

The Great Composers (http://unitedacademymusic.com/ios/#composer) Detailed guide to the lives and works of 227 composers from medieval to the twenty-first century; includes most of the main composers of Western music history; each composer has a high-resolution picture, a detailed biography, and four music video links from YouTube; search the composers through a table in historical order or alphabetical order; an outline of Western music history is included; for music lovers, musicians, music students, and teachers; (iOS 6.0 or later).

The Wagner Files HD (http://www.gebrueder-beetz.de/en/productions/the-wagner-files-app) Explores storytelling with animations, videos, historical letters, music scores, and photographs; the story of Wagner's life is told from Hans von Bülow's point of view starting with the premiere of *Rienzi* in 1842 until Wagner's death in Venice in 1883; forty-one-page graphic novel illustrated by Flavia Scuderi; 240 animated images; more than sixty minutes of Wagner's music; twenty-three interactive points that provide access to historical documents such as letters, photographs, or scores; four modes include Graphic Novel Mode, Animation Mode, Animation Mode with Audio Version, and Auto-Play Mode with Audio Version; music visualization of the overture to *The Flying Dutchman* by Stephen Malinowski; photographs of the most important persons in Wagner's life including Cosima Wagner, Franz Liszt, King Ludwig II, and Hermann Levi; handwritten historical letters and original scores; rare gramophone records of Wagner's operas from the 1920s and 1930s; interviews with Wagner experts Eva Rieger, Oliver Hilmes, Simone Young, Katharina Wagner, and others; interactive map with thirty-three stages and background information; trailer of the film *The Wagner Files*; (iOS 5.0 or later).

Wagner's Ring Cycle App (http://www.naxos.com/catalogue/apps/app .asp?ID=NAPP0701B) App aimed at anyone with an interest in Wagner's monumental operatic achievement *Der Ring des Nibelungen (The Ring of the Nibelung)*; for those acquainted with the cycle or newcomers to it; outline of the work's leitmotifs woven in a sophisticated web throughout all four operas—*Das Rheingold (The Rhinegold)*, *Die Walküre (The Valkyrie)*, *Siegfried*, and *Götterdämmerung (Twilight of the Gods)*; enhanced listening experience; background material; (iOS 5.0 or later).

THE ORCHESTRA APPS

Meet the Orchestra (http://www.meettheorchestra.com) App opens the world of classical music to kids; entertainment and fun interactive experience; app will teach children about the names, appearance, and sounds of the popular orchestra instruments; learn the sections the orchestra is composed of and the musical instruments in each section; differentiate between the instruments both visually and audibly; help develop child's ear as well as his or her musical and general memory; music is one of the main disciplines for finding out the child's potential; it has always been known intuitively, and recently proven scientifically, that early exposure to music is extremely useful for any child's development; along with the instrument learning, children are exposed to some of the best examples of classical music fine-recorded in a professional studio by musicians of a national symphonic orchestra; app is organized into two parts: learning and quizzes; learning part is designed to introduce children to the orchestra and instruments; quiz section develops skills in recognizing instruments by sight and by sound; quiz's difficulty increases with type and level, offering engaging and varying experiences as the learning process progresses; useful for adults in refreshing own knowledge; interesting and educational, simple and effective; includes strings: viola, violin, cello, double bass, and harp; woodwinds: flute, oboe, clarinet, bassoon, and saxophone; brass: French horn, trumpet, trombone, and tuba; percussion: snare drum, bass drum, timpani, cymbals, and piano; (iOS 4.3 or later).

Melody Street HD (http://www.melodystreet.com/apps) *The House On Melody Street* interactive e-book tells the story of the instruments of the orchestra learning to live in harmony; the storyline, written in rhymes and set to a symphonic score, includes lessons about friendship, tolerance, and mutual understanding; musical score represents different dialogues between the various characters, portraying each one with its own unique sound; children are introduced to the musical families of the orchestra including strings, woodwinds, percussion, brass, and pluck, and learn to recognize them by look, sound, character, and family orientation; first chapter introduces the five-story house in which the musical instrument families live; children get to interact and form personal relationships with each of the characters through fun and imagination-filled touch-screen activities; as the story unfolds in the following chapters, users discover that living together is no easy task, when one isn't willing to accept the other's uniqueness; Timpani Tim is too loud, Old Lady Harp's cat goes through the trash, Sammy Snare is rowdy, and Terry Trombone's cooking smells; on one sunny morning, a new neighbor arrives and comes up with a plan to help them live in harmony; listen to the sound of each instrument and follow along with the story; interactive features; fun activities and entertaining musical education; beneficial to the mental, social, and musical development of children; connects children with music and teaches them about the importance of living in harmony; introduction to eighteen musical instruments; child-friendly interactive interface on each page; narrated by Ethan Bortnick; auto-play version plays like a movie; animation; (iOS 3.2 or later).

My First Orchestra (http://www.playtalesbooks.com/en/home) Learn about the different families of musical instruments with an animal band; practice playing the drum, flute, and violin; create own songs; introduce children to music and instruments; (Android 2.2 and up).

My First Orchestra App HD (http://www.naxos.com/catalogue/apps/app .asp?ID=NAPP0102B) Introduction to the orchestra for children aged four and above; the orchestra is a musical team; explore all the instruments in the sequel to Naxos's My First Classical Music App; guide to the orchestra is a little green creature called Tormod, a troll who has come from the top of a mountain in Norway to find out about music; help him find his way home while he shows listeners around the orchestra; every instrument has its own page, with a young performer playing a solo; Tormod shows all of these and sometimes tries the instruments himself; tap any words or pictures and hear the text narrated, extracts of music, animations, and sound effects; many different pieces of music in excellent audio quality; full album of music includes: John Williams: *Superman*, Sarasate: *Carmen Fantasy*, Mozart: Horn Concerto No. 4, Mendelssohn: *The Hebrides (Fingal's Cave)*, Wagner: *Overture to Tannhäuser*, Grieg: *In the Hall of the Mountain King*, and more than thirty more; every instrument in the orchestra is featured with an animated performer; includes beautiful illustrations created specially for the app; friendly, engaging narration; tap on any piece of text to hear it; activities include a quiz on naming the instruments and a go at conducting; full album of music with more than forty different pieces featured throughout the app in excellent recordings; (iOS 4.3 or later).

Sounds of the Orchestra (http://www.smappsoft.com/sounds-of-the-orchestra.html) Learn about the classical musical instruments that constitute the orchestra; wide variety of playing, learning, and listening activities; card matching game to test memory and listening skills by pairing instruments by picture and sound; tap on the soundboard to learn the sounds of each instrument; quiz ear to recognize; explore the four sections of the orchestra; learn about the individual instruments within each section while listening to classical music favorites; for use in any musical education environment; recommended for ages one to seven; (iOS 4.3 or later).

Symphonica (http://www.jp.square-enix.com/symphonica/en/index.html) Music game app; learn what it's like to be an up-and-coming conductor via a unique gaming experience; the story is about Einsatz, the city of music, founded by the legendary conductor Carlos and his companions and brought to where it is today by its citizens, the Audience; talented musicians from around the world flock here with dreams of performing in King Hall at the top of Concerto Tower, which stands tall over the city; Takt has come to Einsatz to pursue his own dream of becoming the world's best conductor; he did not know that he would be in charge of the ramshackle Fayharmonic, an orchestra on the verge of collapse; leaning on Takt as their last hope, the Fayharmonic and their new conductor set their sights on King Hall; thus begins the tale of Takt and his companions in song; lack of specially designated tapping "hot spots" gives user more freedom, allowing them to feel more of what it's like to be a conductor; tempo shifts and real-time, control-linked volume changes; tightly integrated gameplay and story; puts user in the role of a conductor honing the skills of his or her musicians; music game portion is organically integrated with the story; perform at higher score levels to increase orchestra's fame and add new pieces to repertoire; once user clears a given song with a certain number of points, can test skill with a more challenging version of the song; includes many classical masterpieces; features nearly fifty recognizable songs including twenty full performances and approximately thirty rehearsal pieces, including Dvorak's *New World* Symphony and Beethoven's Fifth Symphony; (iOS 5.0 or later).

The Orchestra (http://www.touchpress.com/titles/orchestra/) Esa-Pekka Salonen conducts the world-renowned Philharmonia performing extended extracts from eight works representing three centuries of symphonic music; app allows real-time selection of multiple video and audio tracks, along with an automatically synchronized score and dynamic graphical note-by-note visualization of each piece as it is played; an immersive environment for exploring the music and all the instruments of the orchestra; includes Haydn: Symphony No. 6, Beethoven: Symphony No. 5, Berlioz: Symphonie Fantastique, Debussy: *Prélude à l'après-midi d'un faune*, Mahler: Symphony No. 6, Stravinsky: *The Firebird*, Lutosławski: Concerto for Orchestra, and Salonen: Violin Concerto; features multi-camera, multi-track performances of eight pieces with automatically synchronized full score, curated score, or simplified representation; synchronized BeatMap highlights every note played on a graphical representation of the orchestra; optional audio and/or subtitle commentaries on every piece from both the conductor and the players; discover how each instrument works; read, watch, and hear the musicians themselves explain their role in the orchestra; text on the pieces, composers, and instruments specially written by Mark Swed, music critic of the *LA Times*; interactive video playback allows complete real-time control of the iPad screen in both landscape and portrait orientation; designed for iPad and iPad mini; languages include English, Japanese, German, and French; (iOS 6.0 or later).

JAZZ MUSIC APPS

A Jazzy Day (http://themelodybook.com/a-jazzy-day/) Fun and interactive story that will teach children about jazz music and the instruments typically used to play jazz; features original illustrations and music as well as sounds from real acoustic instruments; music and animation; interactive audio and visual games include "Find the Instrument" or "Which Instrument Sounds Like This?"; tap on the instruments to hear the sounds; interactive pages; touch the animals or instruments and listen; original jazz Big Band music; audio recordings of real acoustic instruments; children will learn to recognize the sight and sound of a saxophone, trumpet, trombone, bass, vibraphone, and more; animated Big Band arrangement demonstrating the use of the different instruments and sections; colorful illustrations; Read to Me: reads and plays the story automatically; Read it Myself: allows user to read the story at own pace; especially designed for little fingers; music instruments featured in the book are: bass, drums, guitar, piano, vibraphone, trumpet, trombone, alto saxophone, baritone saxophone, tenor saxophone, flute, and clarinet; conductor's role; (iOS 5.1 or later).

City Jazz (http://radiocity.me/radioC/pub2.php) Selection of the best radio stations that broadcast continuous jazz in high-quality streaming; (iOS 5.1.1 or later).

Jazzy 123 (http://themelodybook.com/jazzy123/) Learn about music instruments and numbers in a fun and interactive game; (iOS 5.0 or later).

Jazzy ABC (http://themelodybook.com/jazzyabc/) Learn about music instruments and letters in a fun and interactive game; offers a new musical learning experience; learn the alphabet and expand vocabulary through music; interactive pages designed for little fingers; child will learn to recognize a flute, saxophone, trumpet, trombone, bass, vibraphone, drums, and many more instruments by sound and sight; audio recordings of real acoustic instruments from professional musicians; will teach and inspire child to start playing a musical instrument; (iOS 5.0 or later).

Jazzy World Tour (http://themelodybook.com/jazzy-world-tour/) Join two kittens on a magical journey as they travel the world in a hot air balloon; visit Ireland, Japan, Spain, India, and more; explore each country's unique music and culture; play the different musical instruments with interactive fun pages filled with entertaining sounds and vivid illustrations; (iOS 5.1 or later).

The History of Jazz (http://www.955dreams.com/products.html#) Interactive timeline of the history of jazz; made specifically for the iPad; subject is rich with characters and music; traverse the chronological history of jazz as an art form; navigation element is both fun and tactile to multiple touches of fingers; traverse the various genres of jazz; learn about the seminal figures in jazz; legends of jazz are represented with video and images; experience the genre through the artists that made it happen; videos, short biographies, and encyclopedic articles on each artist; list of essential songs and essential albums integrated with iTunes; see the transitions between jazz genres and how the sounds changed; AirPlay support; screensaver mode; introduction to new artists; more than forty-five hours of video content categorized and organized; (iOS 4.2 or later).

POPULAR MUSIC APPS

Band of the Day (http://www.bandofthedayapp.com) Reveals the best new music by delivering one new artist a day, every single day; free music, commercial-free and uninterrupted; user interface, videos, photographs, biographies, and more; discover new favorite bands; listen to today's band, with free full-play songs; explore previously featured bands, similar bands, and the top-rated bands across a wide variety of genres; select the mix tape for a seamless radio experience that highlights some of the best acts in the app; share favorite new discoveries to social networks; available to download for free, worldwide, on any iOS device; listen, discover, share; 100 percent free service worldwide, no country limitations, no subscription required, and no audio ads; access hundreds of previously featured bands and thousands of songs from all over the world, across genres; more than five hours of uninterrupted listening with the Mix Tape, which features more than one hundred of the app's top-rated songs; built-in Airplay support; share new favorites via Facebook, Twitter, and e-mail; (iOS 6.0 or later; Android 4.0 and up).

Billboard Magazine (www.billboard.com) Complete weekly magazine with interactive extras; enhanced format with video, photos, and Billboard's most popular charts playable on Spotify; in-depth analysis of the music business; charts, cover stories, features, special reports, deals, Q&As, album reviews, events, and people; available by subscription; (iOS 5.0 or later).

Digitally Imported (http://www.di.fm/apps) Digital radio network for electronic music fans; listen to more than fifty-five channels of the best electronic music available, each hand-programmed by a channel manager who is an expert in the style; includes Trance, House, Dance, Lounge, Chillout, Techno, Ambient, and many more; styles list; stream music from the app or in the background; control audio and view track titles from the lock screen; save favorite channels; view current and upcoming exclusive radio shows; sleep timer; set data streaming preferences for when using a cellular network; share favorite tracks and channels; optional buffer bar with data display helps user keep track of data usage; (iOS 5.0 or later; Android 2.2 and up).

History of Rock (http://www.history-of-rock-app.com) Comprehensive illustrated history of rock music compiled by renowned rock writer Mark Paytress; includes YouTube videos and iTunes integration; view featured musicians in action or sample song previews by clicking on the relevant icon; discover how rock has grown from its post-war beginnings into a multi-billion dollar industry that caters to all tastes; multimedia app guides the user through rock music's first sixty years, a period that has witnessed many changes as rock continues to draw on ever-wider influences, constantly reinventing itself; from its origins in post-war America to its emergence as the siren call for the late 1960s youth revolution, from the theatrical, troubled 1970s to stadium rock and the digital age, rock music has proven to be resilient and adaptable; explains how and why rock music has become such a force in popular culture; organized in chronological fashion; narrative is divided into eras, each with its own key artists and rock music styles; breakout sections introduce relevant pop culture fads and fashions; timeline provides a continuous thread connecting all the elements that make up the story of rock music; (iOS 5.0 or later).

KandaBi Apps (http://www.kandabisoft.com) KandaBi offers more than one hundred apps featuring a variety of radio stations and programming for numerous genres; (iOS varies by app).

LuceroTech Apps (http://lucerotech.com/games/) Apps include Music Party, Song Trivia, and Guess the Song Game; fun games for music lovers and fans; multiple languages; includes song clips and thousands of popular songs; (iOS varies by app).

M Music & Musicians (https://pixelmags.zendesk.com/home) Entertaining, informative, and exclusive artist-driven features offering insight into the creative process; news and music reviews by the nation's top music journalists; photographs; about music and the people who make it, from songwriters and instrumentalists to producers and other technicians; (iOS 6.0 or later).

MTV Music Meter (http://www.mtv.com/mobile/) Thousands of music videos, tracks, and a selection of photos highlighting artists; available from website only; (iOS 7.0 or later; Android 4.0 and up).

On the Way to Woodstock (http://www.955dreams.com/products.html#) Interactive timeline that explores the phenomenon of how a generation evolved from sock hops to Woodstock; more than one hundred hours of narrative, photography, videos, and music from the 1950s and the 1960s; includes each of the artists that performed at the 1969 Woodstock Art & Music Fair; more than one hundred rare color photos of Woodstock from award-winning photographer Barry Levine; little-known facts about the performers and the events of the festival; experience the essence of life in the 1950s and 1960s featuring family life, social changes, politics, music, and culture; dive deep into the topics and personalities of the time with videos and thoughtful narratives; covers artists from Buddy Holly to Jimi Hendrix, each featured in individual sections filled with photos, videos, and essential songs and albums; includes the characters and artists that chose not to play the festival; AirPlay support for videos; stream audio directly to HiFi speakers; audio/video to Apple TV; more than forty-five hours of video content categorized and organized; screensaver; (iOS 4.2 or later).

The Beatles: The Little Black Songbook (http://www.musicroom.com/music-trivia-apps) The Beatles remain the most critically acclaimed act in the history of popular music; the Beatles had more number one albums on the UK charts and held down the top spot longer than any other musical act in the history of popular music; includes complete lyrics and chords to more than 160 Beatles classics; each song has its own chord library where users can view enhanced details and play audio samples of the chords used; features timeless Beatles songs from all their classic albums including "A Day In The Life" from *Sgt. Pepper's Lonely Hearts Club Band*, "You Never Give Me Your Money" from *Abbey Road*, "Good Day Sunshine" from *Revolver*, "Get Back" from *Let It Be*, "Dear Prudence" from *The White Album*, and many more from their twelve original albums released from 1963 to 1970; contains thirty professionally recorded backing tracks to play along to, including "Drive My Car," "Eight Days A Week," "Help!" "Nowhere Man," "Paperback Writer," "Yesterday," "Ticket To Ride," and "I Saw Her Standing There"; includes a few instructional video lessons to help learn songs like "And Your Bird Can Sing" and "Day Tripper"; app has a built-in tuner so user can play along with the backing tracks and video lessons; fully licensed app; (iOS 3.1 or later).

This Day in Classic Rock (http://www.thisdayinmusicapps.com/classicrock dapp.htm) Includes 366 days of info with thousands of daily rock facts; more than two hundred rock trivia items; more than three hundred quiz questions with an interactive running score; fifty classic rock albums critiqued; seventy-five thousand words of expert critical and historical analysis from AC/DC to the Who and everything in between; free rock ringtone and two free wallpapers; hundreds of rock images drawn from the libraries of *Classic Rock* magazine and the Getty Images photo library; listen to own iPod tracks while reading the notes; direct link to iTunes to buy tracks; links to many rock websites including *Classic Rock* magazine, a site to connect to songs, memorabilia, and more; more than two thousand facts covering all major rock happenings; expand classic rock knowledge at a swipe; interviews with legendary musicians; screensavers; (iOS 5.1 or later).

This Day in Led Zeppelin (http://www.thisdayinmusicapps.com/ ledzeppelinapp.htm) First in a series of apps that chronicle the biggest names in music, including unique audio-visual elements; compiled by the team who run the award-winning This Day in Music website, book, and iPhone app; a celebration of one of the world's biggest and most successful rock acts; lists every gig the band ever played, including set lists, recordings, gigs, and TV performances; daily Zeppelin diary; Zeppelin quiz with hundreds of interactive quiz questions, scored out of ten, with unique score soundclips; includes a detailed critique of every studio track and every Zeppelin album with a unique link to play any Zeppelin track contained in user's iTunes library within the app or buy the track via iTunes; features hundreds of Zeppelin trivia facts with brand-new Zeppelin graphics; comprehensive guide to the group's career, using Zeppelin trivia and facts covering the group's entire history; bonus features include a free rock ringtone; Zeppelin career overview from former *Melody Maker* editor Chris Charlesworth, who spent time with Zeppelin on tour in the United States and flew with the band on their private jet "The Starship" during the 1970s; links to the official Zeppelin stores for all their music, merchandising, and sheet music and the band members' own websites; free Zeppelin art wallpaper; includes many photographs of Zeppelin live and recording venues specially shot for the app; (iOS 4.0 or later).

This Day in Music (http://www.thisdayinmusicapps.com/musicapp.htm) Daily guide to facts, figures, dates, times, places, and the stars that made rock and pop the most vibrant art form of the twentieth and twenty-first centuries; major music happenings 365 days of the year; music quiz with hundreds of questions; what song was number one on the charts on the day user was born; which pop stars were born on the same day as user; hundreds of music trivia facts; companion to the website and book; more than five thousand music facts; hundreds of exclusive images; (iOS 3.1 or later).

This Day in Pink Floyd (http://www.thisdayinmusicapps.com/pinkfloydapp .htm) Contains thousands of music facts covering 366 days of the year; guide to every one of the 167 studio tracks the band officially released; more than one hundred Pink Floyd images; more than two hundred quiz questions; more than two hundred items of trivia; newly restored video of Floyd's 1968 single "Point Me At The Sky"; a free ringtone of "Shine On You Crazy Diamond"; two wallpaper screen images including the *Dark Side* prism and the *Wish You Were Here* cover; play own Pink Floyd music from iTunes library or buy tracks; official links; includes recordings, gigs, and TV performances; the daily Floyd diary; notes to accompany every studio track ever released by Pink Floyd, including non-album singles and B sides; Pink Floyd Quiz has interactive quiz questions with scores out of ten; Pink Floyd trivia; font of Floydian facts in a bumper bundle; link to Pink Floyd's YouTube channel; more than one hundred Floyd images including live shots, backstage shots, and classic Floydian iconography; links to the official Pink Floyd stores for their music, merchandising, and sheet music and the band members' own websites; (iOS 4.0 or later).

This Day in The Rolling Stones (http://www.thisdayinmusicapps.com/ rollingstonesapp.htm) The Rolling Stones' landmark recordings are now embedded in pop culture; celebrating fifty years of Stones history; entries for 366 days of the year; info on all the gigs, tours, TV appearances, recording dates, events, chart positions, album releases, and more; in-depth song notes for fifteen classic Stones albums from the band's debut in 1964 to the classic *Beggars Banquet*, *Sticky Fingers*, *Exile On Main Street*, *Some Girls*, through to the current hits release *GRRR*; more than two hundred Stones songs dissected in more than thirty thousand words; more than three hundred Rolling Stones Quiz questions, scored out of ten; more than three hundred Stones trivia facts; free ringtone and wallpaper with two classic Stones images; more than one hundred Stones images; play own Rolling Stones music in iTunes library or buy tracks; links to websites; (iOS 5.1 or later).

VinylLove (http://www.colormonkey.se/products/vinyllove/index.html) Turns iPad into a turntable that sits on a desk stand; can carry around everywhere; changes the way user experiences music; takes listeners back to a time when they listened to albums one by one, enjoying every single track, in the order they were meant to be listened to; has crackling sound from playing old vinyl records; developer is partners with Binary Peak; (iOS 4.0 or later).

RELIGIOUS MUSIC APPS

Christmas Star Piano!—Learn to Read Music (http://visionsencoded.com/ fun-iphone-apps) Piano app provides visual effects and teaches nine Christmas songs for kids to learn including "Away in a Manger," "Angels We Have Heard on High," "Silent Night," "Oh Little Town of Bethlehem," "Oh Come All Ye Faithful," "Joy to the World," "We Wish You a Merry Christmas," "Oh Come, Oh Come Emmanuel," and "We Three Kings"; pick a song from the rotary dial at the left, then tap the star to see the animation begin; can adjust playback speed by dragging the arrow at the top center of the screen; drag toward the rabbit to speed up the playback or toward the turtle to slow it down or anywhere in between; on the Settings screen, the displayed letter notation can be set to C D E F G A B or Do Re Mi Fa Sol La Ti; (iOS 4.3 or later).

i.Praise Plus (http://www.adventproductions.com) Songs arranged by categories; quick jump to a hymn by number; search engine; save favorites in different playlists; change font styles; supports many different types of songs; saves history of songs navigated; new songs recorded every day; fully functional hymnal player; add favorite songs; (iOS 3.2 or later).

LDS Music (https://www.lds.org/pages/mobileapps?lang=eng) Source for music of the Church of Jesus Christ of Latter-day Saints or LDS, also known as Mormons; browse and search the hymns and children's songbook; read the words and sheet music; listen to accompaniments; English, Spanish, Portuguese, and French are currently supported; preparing additional languages; will be adding additional music, including the Seminary video soundtracks and Young Women's camp songs; (iOS 5.1 or later).

Musical Advent Calendar (http://www.naxos.com/catalogue/apps/app .asp?ID=NAPP0702A) Features twenty-five complete tracks behind numbered doors; each day a new door is unlocked to reveal a new piece of seasonal music from Naxos's classical music catalog; includes Christmas carols and instrumental pieces; old favorites and new discoveries; includes "Ding, Dong, Merrily on High," Vivaldi's *Gloria*, Mozart's *Sleigh Ride, In Dulci Jubilo*, Handel's *Messiah*, and more; (iOS 5.0 or later; Android 2.3.3 and up).

WORLD MUSIC APPS

African Drums (http://www.iplaytones.com) Twenty-six African drums ringtones include: African Baby Lullaby, African DJ Smash, African Drum Smash 1, African Drum Smash 2, African Skies Marimba 1, African Skies Marimba 2, Afro Reggae Groove, Afro Reggae Skunk Groove, Ayub Djembe DJ Fast Groove, Ayub Djembe Fast Groove, Ayub Djembe Groove, Djembe 808 Groove, Djembe 909 Groove, Djembe 909 Groove Hat, Djembe Groove, DJ Fire D-A-N-C-E, Funky Djembe Marimba, Hang Drum, Kalimba, Kalimba Gankogui Bells 1, Kalimba Gankogui Bells 2, Talking Drum, Tongue Drum Blues, Tongue Drum Singing Blues, Udu Seed Mash-Up, and Urban City Mist; (iOS 4.0 or later).

Anghami (http://www.anghami.com) Listen and download music; includes three million Arabic and international songs; search for songs; download songs and listen without the Internet or 3G; build own playlists; discover tunes with personal DJ; share on Instragram, Whatsapp, Twitter, or Facebook; Dolby sound quality; (iOS 6.0 or later; Android 2.3 and up).

ANIME History (AppAge Limited) Japanese Anime hit song collection; (iOS 3.1 or later).

Canadian Musician (http://www.pocketmags.com/information/Default. aspx) Canada's magazine for professional and amateur musicians; published since 1979, the magazine covers prominent Canadian artists, the latest gear, technique, and the business of music; published bi-monthly; features regular columns on guitar, bass, keyboards, percussion, brass, woodwinds, vocal, MIDI, business, songwriting, live sound, recording, and online music; classified ads, opportunities for musicians, new releases, and new products; (iOS 4.3 or later).

Erhu Classic Music (http://www.ndapk.com) The Erhu, also called nanhu or "southern fiddle" and sometimes known in the West as the "Chinese violin" or "Chinese two-string fiddle," is a two-stringed bowed musical instrument used as a solo instrument as well as in small ensembles and large orchestras; Erhu Appreciation and Learning contains Erhu Introduction, Teaching Video, Appreciation of Masterpieces, Save the Recording, and other parts; Erhu Introduction presents the history of Erhu, its construction, variety, characteristic, setting the tune, adjustment, and maintenance; Teaching Video has basic learning guides performed by Professor Chen Chunyan, including Performing Gesture, Holding Gesture, Finger Gesture, Bow-Separation String-Shift, Fast Bow, Cross-Bow String-Shift, and Bow-Shift Skills; Appreciation of Masterpieces virtuoso repertoires with pictures and profiles; one hundred songs; (iOS 4.3 or later).

Hungama—Bollywood Songs & Hindi Music (http://www.hungama.com/#/home) Explore all of Hungama's music library of more than two million songs and music videos spread across Bollywood, Punjabi, Tamil, Telugu, Malayalam, Kannada, Devotional, and much more; for international music lovers; tune in to playlists of favorite songs; Celeb Radio; Discovery Engine plays music based on mood and finds music mix of choice; lyrics and trivia; Bollywood Jukebox; earn points for watching videos; play and share songs, invite friends, or create playlists; Redemption Store where user can redeem points for music downloads, movie tickets, discount vouchers, merchandise, and more; free streaming for music and videos from entire collection of content; sleep mode; create and save playlists; (iOS 5.0 or later; Android 3.1 or up).

Klezmer Melody Book (http://themelodybook.com/klezmer/) App includes traditional Jewish sheet music for voice and piano; download scores and view without requiring an Internet connection; page controls, zoom capabilities, and ability to add comments; (iOS 5.0 or later).

Pow Wow Radio (http://www.powwows.com/2012/08/03/pow-wow-radio-247-native-american-pow-wow-music/) Station with more than two thousand Native American Pow Wow songs; (iOS 4.0 or later).

Rabbi SHALOM (http://www.estebueno.com/blog/rabbi-shaloms-apps-for-iphone-and-ipad/) Rabbi SHALOM performs traditional Jewish dances and songs to sing along; (iOS 4.3 or later).

Top10 of Chinese Traditional Music (http://zkphone.blog.sohu.com) Chinese classical music is a good way to understand Chinese philosophy since Chinese music is influenced by the Eastern philosophy and integrated into the lifestyle of Chinese culture; (iOS 3.0 or later).

Discussion and Activities: Responding and Connecting to Music

*C*reators and performers support interpretations of musical works that reflect expressive intent and incorporate the aesthetic and technical elements of music. The personal evaluation of musical works and performances is accomplished through informed analysis, interpretation, and established criteria. Analyzing social, cultural, and historical contexts and how creators and performers manipulate the elements of music informs the listener's response. Understanding connections to daily life and various contexts enhances how musicians create and perform. Musicians and individuals relate their personal experiences, interests, ideas, and knowledge to how they create, perform, and respond to music.

RESPONDING TO MUSIC TOPICS FOR DISCUSSION

1. Do you feel a piece of music or a painting is more effective in conveying an emotion? Why? What do you feel are the strengths of each medium?
2. How could individuals use music as a tool to improve their home, work, school, or social life?
3. How do individuals choose music to experience?
4. How do the elements of music communicate specific emotions and feelings?
5. How do we discern the expressive intent of musical creators and performers?
6. How do we judge the quality of musical works and performances?
7. How does a rock concert relate to a religious concert? In what ways are they similar? In what ways are they different?
8. How does the title of a piece influence your musical understanding?
9. How does understanding the structure and context of music inform a response?
10. How is music demonstrated in everyday contemporary society? How much music are you exposed to in a single day?
11. In what ways does background music stimulate the senses?
12. Pick an object near where you are sitting. What type of music would you use to represent the object? Why?
13. Think of the last musical theater performance you attended. What kinds of stimuli were used in the production, and how did you react to them? What drew your attention?
14. What draws your attention to the music in a film? What is your emotional response to film music, and what causes that reaction?
15. What experiences have you had with music and the performing arts?
16. What is the value of studying the visual and performing arts?

17. What parts of a musical work appeal to you first? Why?
18. What types of music do you listen to? Why?
19. What would we have left in our society if all art forms, including music, were eliminated?
20. Which is more important to a dance performance, the music or the dance?
21. Which style of music is most appealing to you? Why?
22. Why is the study and understanding of music and aesthetics important for any society?

LEARNING ACTIVITIES

1. Download a tuner and/or metronome app to use when practicing an instrument.
2. Use one of the practice or rehearsal apps to practice playing a piece or singing.
3. Download one of the Internet radio or streaming music apps.
4. Create an iTunes or Google Play playlist.

CONNECTING TO MUSIC TOPICS FOR DISCUSSION

1. Do you feel composers and musicians should feel a sense of social responsibility for the music they create? Why or why not? How do you think the music of today will affect the culture of tomorrow?
2. Every civilization has created music. Why do you think music was an important component of every culture and society? How do you think ancient Romans, cave dwellers, and modern individuals differ or agree in their view of music?
3. How are the technical properties of modern music different from the properties of music from the eras of Baroque or Enlightenment?
4. How do intellectual advancement, social status, and scientific discovery change music?
5. How do musicians make meaningful connections to creating, performing, and responding to music?
6. How do the other arts, other disciplines, contexts, and daily life inform creating, performing, and responding to music?
7. How does music help establish community identity? Provide an example to illustrate your point.
8. How does music help form national identity? Provide an example to illustrate your point.
9. How has music historically been used by political leaders to influence public opinion or to manipulate people? Can you think of a specific example?
10. Much of the music your parents listened to when they were younger is no longer popular today. How does this relate to national, community, or individual identity? Some songs that were popular when your parents were young are still popular today. Why do you think this is the case?
11. Think of an experience in which you used music to serve a social purpose. What was the experience? What social purpose did music serve in that experience?
12. What might a piece of music tell us about its audience, beliefs, themes, ideas, and attitudes of the period in which it was created?
13. What similarities and differences do you see between the use of music in advertising and politics? What characteristics of music allow it to sell a product?
14. Where have you seen music used in religious or spiritual life? Why do you feel music was used in that situation? How did music affect the experience?

15. Why are there so many different forms of sacred music? Do you feel the forms, sounds, and style of sacred music tell us something about the associated belief system?
16. Why do we as a culture include music as a major component of significant ceremonies or events?
17. Why is music important to our community? Why do we invest tax dollars to bring in music to the community? How can you find value in community music?
18. Why was music such a powerful tool in shaping the cultures of New Orleans, Nashville, Detroit, or any other major music center? Do you feel music has contributed to your community's identity? Why or why not?

BAROQUE PERIOD MUSIC TOPICS FOR DISCUSSION

1. How are cantata, mass, and oratorio different from each other? How are they similar?
2. How did Baroque composers and performers use dynamics?
3. In what ways are the cantata, oratorio, and opera similar? In what ways are they different?
4. Is it generally easier to sing or play Baroque melodies? Why?
5. What are the similarities between recitative and aria? What are the differences?
6. What is a fugue? What elements make up a fugue?
7. What is the difference between solo concerto and concerto grosso?
8. What role did women play in creating or performing music of the Baroque period?
9. Why do some individuals claim that opera is the purest integration of the arts?
10. Why do you feel many composers selected the Mass as a subject for musical composition?
11. Why is J. S. Bach considered the most important composer of the Baroque period?
12. Why is opera considered the most characteristic art form of the Baroque period?

CLASSICAL PERIOD MUSIC TOPICS FOR DISCUSSION

1. Social views and patterns are portrayed in music. How did the middle class of the Enlightenment play into the context of music and its performance? How did it contribute to the technical qualities of the music?

2. The theme and variations form was used in many compositions of the Classical period. In what ways could a theme be varied?

3. Was Mozart a successful composer during his lifetime? Why or why not? What types of things did he do to earn a living?

4. What are the differences between the Classical period and the body of music called classical music?

5. What are the similarities and differences between a sonata and sonata-allegro form?

6. What are the similarities between the Baroque and Classical periods? What are the differences?

7. What are two innovations of the Age of Enlightenment that had an effect on the music of the time? What was the effect of each innovation?

8. What changes occurred in the Classical orchestra as compared to the Baroque orchestra?

9. What is sonata-allegro form? How does it affect the music of the Classical period?

10. What is the internal sectional development of the term symphony?

11. Which instrument became a favorite of the Classical period? Why?

12. Who are two important composers of the Classical period? What are some of their contributions to the development of music during this period?

ROMANTIC PERIOD MUSIC TOPICS FOR DISCUSSION

1. How did Robert Schumann's marriage influence his compositions?

2. How did Romantic composers expand the expressive qualities of music?

3. How does the nineteenth-century orchestra differ from the eighteenth-century orchestra?

4. In what ways is the music of the Romantic period similar to other art forms of the period?

5. In what ways were the artists of the Romantic period fascinated with the supernatural and the macabre?

6. What are some characteristics of the Romantic miniature compositions? For which instruments were these compositions usually written?

7. What are some of Richard Wagner's innovations? What were some of the effects of these innovations on later composers and music in general?

8. What are the characteristics and themes of Romantic music as opposed to music created during the Enlightenment?
9. What are the differences between Wagner's operas and Verdi's operas?
10. What are the similarities between Classical and Romantic music? What are the differences?
11. What is nationalism in music? What are some examples of how nationalism can be used in music?
12. What is program music? How is it different from absolute music?
13. What physical problem afflicted Beethoven by about the age of thirty? How did he cope with this problem? How did it change his career?
14. Who are two Romantic composers? What are some of their contributions to music?

CONTEMPORARY PERIOD MUSIC TOPICS FOR DISCUSSION

1. What are some of Stravinsky's musical influences?
2. What are some of the scales used by composers after 1900?
3. What are the differences between music of the twentieth century and previous periods?
4. What is atonal music? What are the characteristics of atonal music? Which composers are most associated with this style?
5. What is impressionism in music? Who is most associated with this style?
6. Who are two composers of the early twentieth century? What were some of their contributions to music?

LEARNING ACTIVITIES

1. Download one of the music appreciation apps and search for an artist, a composer, a composition, or a famous concert or world premiere.
2. Search for apps of favorite artists.
3. Search for apps of favorite genres of music.
4. Complete a music appreciation chart including the following for the Baroque, Classical, Romantic, and Contemporary style periods:
   • List three composers.
   • For each composer, list three compositions.

- List three historical musical events.
- List three historical general events.
- List three artists (painters, sculptors, writers, poets, architects, etc.).
- List three musical innovations.
- List three general innovations.

Bibliography

Axford, Elizabeth C. *Song Sheets to Software—A Guide to Print Music, Software, Instructional Media, and Web Sites for Musicians*. Lanham, MD: Scarecrow Press, 2009.

Biamonte, Nicole. *Pop-Culture Pedagogy in the Music Classroom*. Lanham, MD: Scarecrow Press, 2011.

Bove, Tony. *iPod & iTunes for Dummies*. Foster City, CA: IDG Books Worldwide, 2013.

Burns, Amy M. *Technology Integration in the Elementary Music Classroom*. Milwaukee, WI: Hal Leonard Books, 2008.

Childs, G. W. *Creating Music and Sound for Games*. Boston, MA: Course Technology, 2006.

Collins, Karen. *Playing with Sound: A Theory of Interacting with Sound and Music in Video Games*. Cambridge, MA: MIT Press, 2013.

———. *Game Sound: An Introduction to the History, Theory, and Practice of Video Game Music and Sound Design*. Cambridge, MA: MIT Press, 2008.

Cook, Perry. *Real Sound Synthesis for Interactive Applications (Book and CD-ROM)*. Boca Raton, FL: A K Peters/CRC Press, 2002.

Foreman, Greg, and Kyle Pace. *Integrating Technology with Music Instruction*. Van Nuys, CA: Alfred Publishing, 2008.

Freedman, Barbara. *Teaching Music through Composition—A Curriculum Using Technology*. New York, NY: Oxford University Press, 2013.

Gallagher, Mitch. *The Music Tech Dictionary*. Boston, MA: Course Technology PTR, 2009.

Gordon, Steve. *The Future of the Music Business: How to Succeed with the New Digital Technologies*. Milwaukee, WI: Hal Leonard Books, 2014.

Harvell, Ben. *Make Music with Your iPad*. Indianapolis, IN: John Wiley & Sons, 2012.

Hoffert, Paul. *Music for New Media: Composing for Videogames, Web Sites, Presentations and Other Interactive Media (Book & CD)*. Boston, MA: Berklee Press, 2007.

Huber, David Miles, and Robert E. Runstein. *Modern Recording Techniques*. Burlington, MA: Focal Press, 2013.

Jenkins, Mark. *iPad Music in the Studio and on Stage.* Burlington, MA: Focal Press, 2013.

Kohn, Al, and Bob Kohn. *Kohn on Music Licensing.* Frederick, MD: Aspen Publishers, 2009.

Kompanek, Sonny. *From Score to Screen: Sequencers, Scores, and Second Thoughts: The New Film Scoring Process.* New York, NY: Schirmer Books, 2004.

Kusek, David. *The Future of Music: Manifesto for the Digital Music Revolution.* Boston, MA: Berklee Press, 2005.

Marks, Aaron. *The Complete Guide to Game Audio: For Composers, Musicians, Sound Designers, and Game Developers.* Burlington, MA: Focal Press, 2008.

Middleton, Chris. *Creating Digital Music and Sound: An Inspirational Introduction for Musicians, Web Designers, Animators, Videomakers, and Game Designers.* Burlington, MA: Focal Press, 2006.

Minute Help Guides. *A Newbies Guide to Using GarageBand for the iPad.* Anaheim, CA: Minute Help Press, 2012.

NAMM. *International Music Market Show Directory.* Carlsbad, CA: NAMM, 2014.

Phillips, Winifred. *A Composer's Guide to Game Music.* Cambridge, MA: MIT Press, 2014.

Read, Gardner. *Music Notation: A Manual of Modern Practice.* New York, NY: Taplinger, 1979.

Rothermich, Edgar. *GarageBand for iPad.* North Charleston, SC: CreateSpace Independent Publishing Platform, 2014.

———. *Logic Remote (iPad).* North Charleston, SC: CreateSpace Independent Publishing Platform, 2013.

Rudolph, Thomas. *Teaching Music with Technology*, 2nd ed. Chicago, IL: GIA Publications, 2005.

Rudolph, Thomas, and James Frankel. *YouTube in Music Education.* Milwaukee, WI: Hal Leonard Books, 2009.

Rudolph, Thomas, and Vincent Leonard. *The iPad in the Music Studio.* Milwaukee, WI: Hal Leonard Books, 2014.

———. *Musical iPad.* Milwaukee, WI: Hal Leonard Books, 2013.

Rudolph, Thomas, Floyd Richmond, David Mash, and Peter Webster. *Technology Strategies for Music Education.* Wyncote, PA: TI:ME Publications, 2005.

Sexton, Jamie. *Music, Sound and Multimedia: From the Live to the Virtual.* Edinburgh, Scotland, UK: Edinburgh University Press, 2007.

TI:ME. *Technology Guide for Music Educators.* Milwaukee, WI: Hal Leonard Books, 2005.

Viss, Leila J. *The iPad Piano Studio.* Buena Park, CA: Webgines Publishing, 2013.

Watson, Scott. *Using Technology to Unlock Musical Creativity.* New York: Oxford University Press, 2011.

Williams, David, and Peter Webster. *Experiencing Music Technology*, 3rd ed. New York: Schirmer Books, 2008.

Zager, Michael. *Writing Music for Television and Radio Commercials and More.* Lanham, MD: Scarecrow Press, 2008.

———. *Music Production: A Manual for Producers, Composers, Arrangers, and Students.* Lanham, MD: Scarecrow Press, 2006.

WEBSITES

AfterDawn Glossary of Tech Terms iOS. http://www.afterdawn.com/glossary/term
.cfm/ios
Android A to Z: An Android Glossary. http://www.androidcentral.com/android-
glossary
The Android Dictionary. http://www.androidcentral.com/dictionary
Android OS Help. https://support.google.com/android/?hl=en
Apple iOS Terms Glossary. http://www.gsmarena.com/glossary.php3?term=apple-ios
Apple Manuals. http://support.apple.com/manuals/
Glossary of Android OS Terms. http://www.androidglossary.com
iPhone and iPad Glossary. http://www.imore.com/glossary
Tech Terms Android. http://www.techterms.com/definition/android
Tech Terms iOS. http://www.techterms.com/definition/ios

About the Author

Elizabeth C. Axford has a master of arts degree in music from San Diego State University and a bachelor of arts degree in music from the University of Illinois, Urbana-Champaign. She currently facilitates online music and arts courses for the University of Phoenix. During her thirty years of teaching studio piano and voice, Liz has held numerous recitals with her students. Her teaching studio, Piano Press Studio (www.pianopress.com), received the Best of Del Mar Award seven years in a row from 2008 to 2014. Liz is an award-winning songwriter and pianist with over 2,500 usages of her tracks in films, TV shows, videos, commercials, bumpers, promos, and new media projects. Her music can be heard on artists' CDs, digital compilations, websites, video games, apps, ringtones, greeting cards, gift bags and collections, and in musical theater productions. Liz is an active member of ASCAP, NARAS, NSAI, NAMM, MTNA, CAPMT, and MTAC, and has been a guest speaker and/or exhibitor at numerous state and national conferences. She has voted on and attended the Grammys since 2000, as well as numerous industry events in Los Angeles, Nashville, and Miami since 1989. During her twenty-year tenure as Nashville Songwriters Association International (NSAI) Regional Workshop Coordinator in Miami, Florida, and San Diego, California, Liz was responsible for more than three hundred monthly workshops, songwriter showcases, and special events. She has received service awards from NSAI, numerous song-writing awards, and six ASCAP PLUS Awards. Liz is the author of *Traditional World Music Influences in Contemporary Solo Piano Literature* and *Song Sheets to Software—A Guide to Print Music, Software, Instructional Media, and Web Sites for Musicians*, published by Scarecrow Press.

CPSIA information can be obtained
at www.ICGtesting.com
Printed in the USA
LVHW04s1557240718
584775LV00011B/721/P